Advanced Decision Making for HVAC Engineers

Javad Khazaii

Advanced Decision Making for HVAC Engineers

Creating Energy Efficient Smart Buildings

 Springer

Javad Khazaii
Engineering Department
Kennesaw State University (Marietta Campus)
Marietta, GA, USA

ISBN 978-3-319-81486-5   ISBN 978-3-319-33328-1   (eBook)
DOI 10.1007/978-3-319-33328-1

Printed on acid-free paper

This Springer imprint is published by Springer Nature
The registered company is Springer International Publishing AG Switzerland

To
Love of My Life Hengameh;

also to
My lovely mother Efat and distinguished
brother Dr. Ali.

Dad, I've missed you!

Preface

Every architect or engineer in his daily work routine faces different complicated problems. Problems such as which material to specify, which system to select, what controls algorithms to define, what is the most energy efficient solution for the building design and which aspect of his project he should focus on more. Also, managers in architectural and engineering firms, on a daily basis, face complicated decisions such as what project to assign to which team, which project to pursue, how to allocate time to each project to make the best overall results, which new tools and methods to adopt, etc. Proper decision making is probably the single most important element in running a successful business, choosing a successful strategy and confronting any other problem. It is even more important when one is dealing with complicated architectural and engineering problems. A proficient decision maker can turn any design decision into a successful one, and any choice selection into a promising opportunity to satisfy the targets of the problem question to the fullest. The paradigm of decision theory is generally divided into two main sub-categories. Descriptive and normative decision making methods are the focuses of behavioural and engineering type sciences respectively. Since our focus in this book is on HVAC and the energy engineering side of the decision making, our discussions are pointed at normative type decision making and its associated tools to assist decision makers in making proper decisions in this field. This is done by applying available tools in decision making processes commonly known as decision analysis. Three generally accepted sub-categories of decision analysis are decision making under uncertainty, multi-criteria decision making and decision support systems. In the second half of this book I will attempt to present not only a brief explanation of these three sub-categories and to single out some of the most useful and advanced methods from each of these sub-categories, but I will also try to put each of these advanced techniques in perspective by showing building, HVAC and energy related applications in design, control and management of each of these tools. Finally, I will depict the smart buildings of the future which should be capable of autonomously executing these algorithms in order to operate efficiently and intelligently, which is categorically different from the buildings that are currently and unsophisticatedly called as such. As it has always been my strategy and similar to my previous work, I have made my maximum effort not to bore the readers with too many details and general descriptions that the reader can find in many available sources in which each method has been expressed in depth. To the contrary I try to explain the subjects briefly but with enough depth to draw the reader's desire and attention towards the possibilities that these methods and tools can generate for any responsible architect, HVAC and energy engineer. I have provided numerous resources for studying the basics of each method in depth if the reader becomes interested and thinks he can conjugate his own building related problems with either one of these supporting methods for decision making. Even though the material explained in the second half of the book can be very helpful for any decision maker in any field, obviously my main targeted audience are the young architects and HVAC engineers and students that I hope to expose to the huge opportunities in this field. The goal is to make them interested in the topic and give them the preliminary knowledge to pursue the perfection of the methods and in this path advance the field in the right direction and with the most advanced available methods.

In order to be able to describe these opportunities in the second part of the book, I have dedicated the first part of this book to a general and brief review of the basics of heat transfer science which have a major role in understanding HVAC and energy issues, load calculations methods and deterministic energy modelling, which are the basic tools towards understanding the energy consumption needs of the buildings and also more importantly some of the highest energy consuming applications in building, HVAC and energy engineering. This will help the reader to quickly refresh his knowledge about the basic heat transfer concepts which are the elementary required knowledge to understand HVAC and energy topics, learn more about the essentials of load calculations methods and deterministic available energy modelling tools in the market and of course learn about the big opportunities in high energy consuming applications for utilization of the described decision making tools in order to save energy as much as possible.

The most important challenge in the next few decades for our generation is to generate enough clean energy to satisfy the needs of the growing population of the world. Our buildings and their systems consume a large chunk of this energy and therefore this fact positions us as architects and engineers in the centre of this challenge. We will not be able to keep up with this tremendous responsibility if we cannot make the correct decisions in our design approach. It is therefore the most basic necessity for any architect, HVAC and energy engineer to familiarize himself not only with the knowledge of his trade but also with the best available decision making tools. I hope this book can help this community to get themselves more familiar with some of the most advanced methods of decision making in order to design the best and most energy efficient buildings.

Marietta, GA, USA   Javad Khazaii

Acknowledgement

I would like to thank my father that is always in my memory for all he did for me. I also want to thank my brother Dr. Ali Khazaei, my friend Dr. Reza Jazar and my mentor Professor Godfried Augenbroe for their deep impacts on my scientific achievements, and my Mom and my Wife for their endless love and support.

Furthermore I wish to extend my additional appreciation towards Dr. Ali Khazaei for his continuous involvement in discussing, debating, and commenting on different material presented in the book during the past 2 years without whose input and help I could not be capable of completing this work.

Javad

Contents

About the Author

Dr. Javad Khazaii holds a B.Sc. in Mechanical Engineering from Isfahan University of Technology, an MBA with a concentration in computer information systems form Georgia State University and a PhD in Architecture with a major in building technology and a minor in building construction from Georgia Institute of Technology. He is a registered engineer and a LEED accredited professional with more than two decades of professional project management, design and energy modelling experience. He has been an adjunct faculty in the engineering department of Kennesaw State University (previously known as Southern Polytechnic State University) since early 2011.

Dr. Khazaii has co-authored scientific articles and conference proceedings for the ASME and IBPSA, and he was one of the original contributors to the State of Qatar (Energy) Sustainability Assessment System (QSAS). His team was awarded first place in the International Building Performance Simulation Association's (IBPSA) annual group competition while he was working on completing his PhD degree.

His first book "Energy Efficient HVAC Design, An Essential Guide for Sustainable Building" was published by Springer in 2014 and immediately charged to the Springer's best sellers list in "Energy" category.

Introduction

Abstract: Behind any success story or any major achievement one can find traces of one or a series of properly made decisions. Success stories such as designing a state of the art building and achievements such as participating in saving the scarce energy resources for future generations while designing that state of the art building are not simple tasks and cannot be accomplished easily. In this context, subjects that the architects and engineers in general and HVAC and energy engineers in particular are commonly occupied with are extremely important. In the United States alone about 40 % of the energy and close to 70 % of the electricity produced are consumed by buildings. Therefore what the architects and engineers do, if not properly done, can cause dire consequences not only for us but also for future generations.

Keywords: Energy, Efficient, Sustainable, Knowledge, Holistic, Evolutionary, energy consumption, healthcare facilities, data centres, cleanrooms, laboratories, decision making under uncertainty, fuzzy logic, Pareto optimality, genetic algorithm multi-objective optimization, artificial neural network, game theory, buildings of the future, smart buildings

Behind any success story or any major achievement one can find traces of one or a series of properly made decisions. Success stories such as designing a state of the art building and achievements such as participating in saving the scarce energy resources for future generations while designing that state of the art building are not simple tasks and cannot be accomplished easily. In this context, subjects that the architects and engineers in general and HVAC and energy engineers in particular are commonly occupied with are extremely important. In the United States alone about 40 % of the energy and close to 70 % of the electricity produced are consumed by buildings. Therefore what the architects and engineers do, if not properly done, can cause dire consequences not only for us but also for future generations.

In my previous book "Energy Efficient HVAC Design, An Essential Guide for Sustainable Building" I did my best to briefly introduce most of the important and necessary topics that I thought are essential knowledge for an informed HVAC or architectural engineering student or young professional. In this new book my intention is not only to build even more on advancing the related and important topics for this community of practice, but also to introduce the most advanced tools and techniques in order to help a well-informed engineer select the best choices among all possible design options.

This book is divided into four main sections. In the first section I have presented a short, yet holistic summary of the general heat transfer topics. No other basic science has more roots and relevancy in HVAC and energy engineering than heat transfer. I have tried to shine some light on and remind the readers of the basic concepts of conduction, convection and radiation. Each of these heat transfer modes are responsible for parts of the overall heat gain or heat loss in the building which are the main target of the HVAC, energy and architectural engineering professionals to control in order to build a safe and comfortable space for our specific functions.

I have dedicated the rest of section one to the building load calculation and energy modelling basics. A short discussion about the evolutionary map and therefore different proposed methods of load calculation in the past few decades have been presented which shows the importance and role of heat transfer laws in defining the heat gain by buildings for the worst condition and therefore system selection. More discussion has been presented which is directed towards how to perform energy modelling in order to calculate the yearly energy consumption of the building based on the current guidelines presented by the energy standards.

In the second section I have presented a discussion about a few high energy consuming applications for building HVAC systems such as healthcare facilities, data centres, cleanrooms and laboratories. Discussion is directed towards representing the fact that designing these applications are not only a daunting task from the point of view of complexity and control of such large and diverse loads, but also to represent the huge opportunities for saving energy if we design more intellectually and find better solutions for our routine design approaches.

In the third section which is also the lengthiest section of the book, I have focused on a more general and yet extremely important discussion. First, I call it "general" because even though the examples and writings are generally directed towards the HVAC and energy consumption solutions, at the same time utilizing these methods and techniques are as good for any other field as they are for the HVAC and energy field of engineering. Also, I call it "extremely important" because in this section I have reviewed and presented some applications for some of the most advanced techniques in decision making that, even though they are relatively new but known in the world of academics, most of the professionals and their firms are not in general familiar with and therefore at best—to be generous—rarely use them in their daily design and managerial routines. Adopting any of these techniques in engineering routine works and design processes can turn the page on how we design and how we make important decisions in a way that was not possible before. In this section I have provided a general overview of the decision theory with the main target of describing available decision analysis tools. I have briefly described some of the available advanced decision making methods such as decision making under uncertainty, fuzzy logic, Pareto optimality, genetic algorithm multi-objective optimization, artificial neural network and game theory and then used them for solving some hypothetically HVAC and energy engineering applications.

In the final section of the book I have taken the concept of smart decision making even further by imagining the fact that in the near future buildings on their own will be capable of making decisions more accurately and in a more-timely manner—by utilizing these decision making methods—than we are doing in the current environment. This is a true definition of buildings of the future and smart buildings.

My main objective in writing this book, similar to my previous book, was to expose the students and young professionals to a combination of the academic and professional topics. Topics that, based on the students pursuance of graduate education or choosing professional life, they may or may not be exposed to in either path. My belief is that our field of engineering will not advance as it should if the professionals continue working in a vacuum without understanding what the advanced related topics are, and academic researchers will not have a realistic picture of the HVAC and energy universe without understanding what the real world problems and opportunities are. Getting both of these groups to understand each other's issues and talk with the same language is what I think can make this field move as fast as it should in a correct route.

My other important objective in writing this book was to inspire the young professionals and students to defy the conventional way of thinking about how to design and conserve energy in our field of work and encourage them to make design related decisions in a way that scientifically should be made. Such decisions should be made with proper tools and methods—either those that currently exist or those that are to be developed by the new generation of engineers themselves.

At this point I need to emphasize that material provided in this book does not supersede the requirements given in any local and national code, regulation, guideline and standard, and should not be regarded as a substitute for the need to obtain specific professional advice for any particular system or application. The main purpose of the book is to derive attention of the interested young readers to some available advanced techniques that could be mastered by the readers through hard study and work which then may help him or her to make better decisions while designing buildings. Some hypothetical examples and a brief description of each method have been given and I did my best to give some good sources of study about each topic. It is the reader's responsibility to familiarize himself or herself with the subjects and methods through available books and other resources that discuss and teach the methods in depth before deciding to implement each method to help his decision making process. Each better decision that we can make regarding energy consumption and design can help all of us to have a better planet and future.

Part I
Basics of Heat Transfer, Load Calculation, and Energy Modeling

Chapter 1
Heat Transfer in a Nutshell

Abstract No other fundamental science has more relevancy to the topics of energy, building and its HVAC systems than heat transfer. We use basics of convective, radiative, and conductive heat transfer to calculate the amount of heat or cool imposed by the exterior environment and internal elements on our buildings. We use basics of all the three types of heat transfer to calculate how much of this heat or cool is going to be transferred to inside air or out of our buildings. We use conduction mode laws to size our insulations inside the walls of our buildings and around our equipment to reduce the unwanted heat transfer from or to our buildings and systems. We add (deduct) the internal heat load which is transferred to the space in form of part radiation (sensible portion) and part convection to our heat gain (loss) calculations.

Keywords Heat transfer • Conduction • Convection • Thermal radiation • Heat exchanger • Fourier's Law • Boltzmann constant • Black surface • Extraterrestrial solar flux • Reynolds number • Prandtl number • Rayleigh number • Heat transfer coefficient

"The science of heat transfer seeks not merely to explain how heat energy may be transferred, but also to predict rate at which the exchange will take place under certain specified conditions. Heat transfer supplements the first and second principles of thermodynamics by providing additional experimental rules that may be used to establish energy-transfer rates" [1]. No other fundamental science has more relevancy to the topics of energy, building and its HVAC systems than heat transfer. We use basics of convective, radiative, and conductive heat transfer to calculate the amount of heat or cool imposed by the exterior environment and internal elements on our buildings. We use basics of all the three types of heat transfer to calculate how much of this heat or cool is going to be transferred to inside air or out of our buildings. We use conduction mode laws to size our insulations inside the walls of our buildings and around our equipment to reduce the unwanted heat transfer from or to our buildings and systems. We add (deduct) the internal heat load which is transferred to the space in form of part radiation (sensible portion) and part convection to our heat gain (loss) calculations. We also add the latent heat that is converted to heat gain by convection rules to the calculated external heat or cool to come up with our final heating or cooling demand to condition our buildings. The

© Springer International Publishing Switzerland 2016
J. Khazaii, *Advanced Decision Making for HVAC Engineers*,
DOI 10.1007/978-3-319-33328-1_1

heat transfer process in our buildings continues as the generated heat and cool in the
building plant moves via for example chilled water and hot water to the building
inside air through the heating and cooling coils conduction and of course forced
convective act of air handling unit fans. Heat transfer is the science that directly
deals with calculations for heat exchanger design in depth which is in the heart of
the building plant for HVAC systems and includes all types of heat transfer modes.
We insulate our ducts and pipes to prevent the heat transfer through the pipes and
ducts in forms of conduction, convection, and radiation. Inside the chiller plant the
heat that is extracted from the inside air space by the chilled water will be rejected
to the outside air by heat transfer process. Therefore, designing the HVAC system
for a building and analyzing the energy consumption of the building would not be
possible without a good understanding of the concepts and methods of heat transfer.
All these justify dedicating a short passage of this book for refreshing our knowledge about the basic modes of heat transfer at this point. In the next few paragraphs
I review the main three modes of heat transfer which are conduction, convection,
and radiation as a refresher short discussion.

Conduction

Building shape is defined by its envelope which separates its interior from the
outdoor environment. The main purpose of air conditioning system is to keep a
comfortable and functional interior space in which people can live and work
without being exposed to the extreme outdoor conditions. Building envelope
usually consists of two main elements of opaque and glazing systems. The opaque
system consists of exterior walls, doors, frames, and roofs, and glazing systems
when installed on the walls are known as windows and when installed on the roofs
are called sky-lights. The majority of the heat transfer through the opaque parts of
the building is based on conductive heat transfer mode due to the temperature
gradient between inside and outside surfaces temperature. This heat transfer happens from surface with higher temperature towards the surface with lower temperature. "Conduction may be viewed as the transfer of energy from the more energetic
to the less energetic particles of a substance due to interactions between the
particles" [2]. The exterior surface of the building specifically in summer not
only carries the transferred convective heat which exists in the outdoor environment, but also and due to the fact of absorption of both direct and indirect sun
radiation carries the heat which stores in the wall. This increases wall temperature
with of course some time delay as well.

The effect of conductive heat transfer specifically during the summer time is
much less than radiative heat transfer through the glazing system, which is
discussed a little later in the following sections.

In a simplified steady state one dimensional, without heat generation inside,
conductive heat transfer inside a body is calculated by using Fourier's Law of heat

conduction which states that the energy transferred through a body is proportional to the normal temperature gradient and is calculated via using the following equation:

$$q_x = -K.A.\partial T/\partial x \qquad (1.1)$$

In this equation q_x is the heat transfer rate with its dimensions in SI and English system equal to W (Watts) and Btu/h (British Thermal Unit per hours) respectively and $\partial T/\partial x$ is the gradient of temperature in direction of heat flow in °C/m or °F/ft, while A is the representative of the conduction area in m^2 or ft^2, and K is the conductive heat transfer coefficient. The conductive heat transfer coefficient dimensions in SI and English systems are W/m°C and Btu/h ft°F respectively. Based on type of the material, thermal conductivity value can change drastically from one material to another. Of course the negative sign in Fourier's equation is used to off-set the negative value of temperature difference that is flowing from higher to lower temperature surface. K is measured based on experiment and is listed in different references for different materials.

For a simple wall thickness of Δx and surface temperatures of T_2 and T_1, Eq. 1.1 can be rewritten in the following form:

$$q_x = -K.A.(T_2 - T_1)/\Delta x \qquad (1.2)$$

If Eq. 1.2 is reformatted in the following form:

$$q_x = (T_1 - T_2)/(\Delta x/(K.A)) \qquad (1.3)$$

it can be compared to the electrical circuits law where $(T_1 - T_2)$ acts similar to voltage change, q_x acts similar to the electricity current, and $(\Delta x/K A)$ acts similar to resistance. Further if the wall is built from multiple layers of different materials with different thicknesses and K values, the equation that can represent the conductive heat transfer through the whole wall can be described (e.g., for a three-layer wall shown in Fig. 1.1 below) with Eq. 1.4.

$$q_x = (T_1 - T_4)/((\Delta x1/(K_1.A)) + (\Delta x_2/(K_2.A)) + (\Delta x_3/(K_3.A))) \qquad (1.4)$$

where T_1 and T_4 represent the temperatures of two outer and inner surfaces of the wall. The layers can be made of material, insulations, and air gaps).

Of course we always have to consider the convective effects of air movement outside and inside the building on two sides of the wall that would generate additional resistance in the path of the heat transfer. Convective heat transfer adjacent to a wall or roof will follow the general Newton's law of (convective) cooling which I discuss in further detail later in this chapter:

$$q = h.A.(T_s - T_\infty) \qquad (1.5)$$

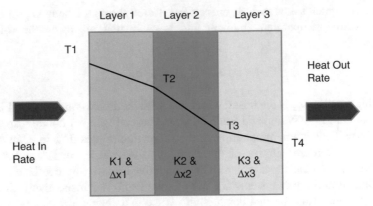

Fig. 1.1 Conduction through multilayer wall

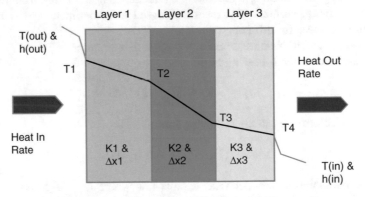

Fig. 1.2 Heat transfer (conduction and convection) through a multilayer wall

with q representative of convective heat transfer rate between the wall surface and air in units of W or Btu/h, h representative of convective heat transfer coefficient in units of W/ $m^{2\circ}$C or Btu/h. $ft^{2\circ}$F, and A representative of heat transfer area in units of m^2 or ft^2. Finally T_s and T_∞ are wall surface and air temperatures. Reformatting Eq. 1.5 similar to what we do with Eq. 1.2 to compare it to electrical circuits law results in the following arrangement:

$$q = (T_s - T_\infty)/(1/(h.A)) \tag{1.6}$$

with $(1/(h.A))$ representative of convective resistance in the equation. Combining resistance from interior and exterior heat convection with the conductive resistance of for example our three-layer wall we have Fig. 1.2 and can derive Eq. 1.7 below:

$$q_x = (T_{outside} - T_{inside})/((1/(h_{outside}.A)) + (\Delta x_1/(K_1.A)) + (\Delta x_2/(K_2.A))$$
$$+ (\Delta x_3/(K_3.A)) + (1/(h_{inside}.A)))$$

$$(1.7)$$

with $((1/(h_{outside}.A)) + (\Delta x_1/(K_1.A)) + (\Delta x_2/(K_2.A)) + (\Delta x_3/(K_3.A)) + (1/(h_{inside}.A)))$ representative of overall heat transfer resistance **R** between outside and indoor air. It is common in HVAC applications to simplify Eq. 1.7 into:

$$q = U.A .\Delta T_{overall} \qquad (1.8)$$

with $U(= 1/(\mathbf{R}.A))$ defined as overall heat transfer coefficient (for our example) equal to $(1/((1/h_{outside}) + (\Delta x_1/K_1) + (\Delta x_2/K_2) + (\Delta x_3/K_3) + (1/h_{inside})))$, A representative of heat transfer area perpendicular to the direction of heat transfer and $\Delta T_{overall}$ representative of temperature difference between inside and outside air. Different U values in W/ m$^{2\circ}$C or Btu/h. ft$^{2\circ}$F should be calculated based on thickness and heat resistance values of construction material and convective heat resistance inside and outside the building. That should be used as part of envelope conductive heat loss or gain calculations. Similar equations can be written for the conductive heat transfer through roofs and also overall heat transfer coefficient between the outside and indoor air on each side of the roof. The conductive resistance along with shading coefficient of glazing system usually should be collected from the manufacturer's data sheet for conductive and radiative portion of heat transfer through the building glazing.

Thermal Radiation

All different materials since their surface temperature is above the absolute zero emit thermal radiation. "Most of the heat that reaches you when you sit in front of a fire is radiant energy. Radiant energy browns your toast in an electric toaster and it warms you when you walk in the sun" [3]. Stefan Boltzmann showed the relation between thermal radiation and absolute temperature as follows:

$$(q/A) = \sigma T^4 \qquad (1.9)$$

Where units of q is in W, A is in m^2, T is in K, and σ (Boltzmann constant) is 5.670×10^{-8} W/m^2 K^4.

Thermal radiation is emitted in the form of electromagnetic waves and does not need material medium to transfer. Most of thermal radiation happens between 0.1 and 10 μm, where only the section between 0.4 and 0.7 μm is visible light [1]. Materials also can absorb, transmit, or reflect received radiation emitted by its surrounding, that follows the following equation.

$$\alpha + \rho + \tau = 1 \tag{1.10}$$

Where α is the absorbed portion, ρ is reflected fraction, and τ is the transferred percent of total emitted radiation received by the surface. When the surface is opaque, there will be no transmission and therefore the equation will be reduced to:

$$\alpha + \rho = 1 \tag{1.11}$$

Both the above equations are valid for any specific wavelength as well. As a result we can write:

$$\alpha_\lambda + \rho_\lambda + \tau_\lambda = 1 \tag{1.12}$$

and

$$\alpha_\lambda + \rho_\lambda = 1 \tag{1.13}$$

Another important definition when studying thermal radiation is the black surface (body). A black surface is an ideal surface that not only absorbs all the emitted radiation regardless of the direction and wavelength of the radiation on it, but also for any given wavelength and temperature emits the maximum amount of energy. Total emissive power of a black surface is called e_b, which is the radiant flux emitted from the surface of the black body in units of W/ m^2. The emissivity of other non-black surfaces is defined as:

$$\varepsilon = \ e / e_b \big| \text{ at same temperature} \tag{1.14}$$

where $0 \leq \varepsilon \leq 1$.

If we assume a body is trapped inside a black enclosure and exchanges energy with this enclosure as it receives radiant flux(f) from the enclosure. At equilibrium the energy absorbed by the body shall be equal to the energy emitted by the black enclosure. Therefore, for the body at equilibrium we will have [1]:

$$e .A = q .A .\alpha \tag{1.15}$$

Now assume the body inside the enclosure is a black body. Since α (absorptivity) of a black body is equal to 1, the above equation will be reduced to [1]:

$$e_b. A = q .A . 1 \tag{1.16}$$

Dividing the above Eq. 1.15 over Eq. 1.16 will result in the following equation:

$$e/e_b = \alpha \tag{1.17}$$

Comparing the above equation (Eq. 1.17) with Eq. 1.14 results in an important outcome of $\alpha = \varepsilon$, for a black surface [1].

Emissive power of black surface can be calculated using Planck's law as follows: (other formats of this equations have been presented by different sources as well)

$$e_{b\lambda} = 2\,\pi\,C_1/\lambda^5 \left[\exp\left(C_2/\lambda\,T\right)^{-1}\right] \qquad (1.18)$$

where $C_1 = 0.596 \times 10^{-16}$ W m^2, $C_2 = 0.014387$ W K, λ is the wavelength in micron, and T is the absolute temperature in K. Temperature and wavelength in which the maximum emissive power for a black surface happens are related by Wien's law as follows:

$$\lambda_m.\,T = 0.00290 \text{ m. }^\circ K \qquad (1.19)$$

The effects of radiation in building load calculation appears in a couple of places. First total solar radiation incident on the exterior surfaces of receiving walls and roofs, and the emitted radiation from these surface back to the sky are used to calculate sol-air temperature. This temperature then will be used as the overall external wall or roof surface temperature to calculate the heat conduction through walls and roofs. Sol-air temperature is defined by the following equation in which E_t represents total clear sky irradiance incident on surface and ΔR represents the difference between long-wave radiation incident on the surface from the sky and surroundings and radiation emitted by the black body at outdoor temperature, and $h_{outside}$ is the convectional heat transfer coefficient of the outdoor air [6].

$$T_{\text{sol-air}} = T_{\text{outside}} + \alpha\,E_t/h_{\text{outside}} - \varepsilon\,\Delta R/h_{\text{outside}} \qquad (1.20)$$

Total clear sky irradiance reaching the glazing system "E_t" is the next place of appearance of radiation in building calculations. Let us see how we can analyze the radiation effect of Sun on our design buildings based on the presented approach by ASHRAE Fundamentals Handbook. Total value of solar energy flux of Sun (Extra-terrestrial Solar Flux) just outside the Earth's atmosphere is known to swing between two lower and upper limits of 419–447 Btu/h ft^2 depending on the time of the year. This flux after entering the Earth's atmosphere and according to its traveling path—that is if it is moving directly towards the ground or being diffused by the pollution and similar particles in the Earth's atmosphere before reaching the ground—divides to two portions of beam normal irradiance and diffused horizontal irradiance. Beam and diffused irradiances are each functions of extraterrestrial solar flux, air mass (a function of solar altitude angle), and clear sky optical depth of beam and diffused irradiances. From the point of view of a receiving surface such as a wall, a window or a roof, a portion of this sun flux either hits this surface directly or being redirected towards this surface after hitting the ground or the adjacent surfaces. Therefore, total clear sky irradiance "E_t" reaching the receiving surface will contain beam, diffused, and ground reflected components.

The beam component of the clear sky total irradiance reaching the receiving surface on the ground is a function of the beam normal irradiance and solar incident angle, while the diffused component of it is a function of diffused horizontal irradiance and solar incident angle. On the other hand, the ground reflected component of the clear sky total irradiance reaching the receiving surface is a function of both beam normal and diffused horizontal irradiance, ground reflectance, surface tilt angle, and solar altitude angle as well. As mentioned earlier a sum of these three components (clear sky total irradiance) is used to calculate the opaque surface sol-air temperature in order to calculate the heat conduction through them. The beam component alone will be used to calculate the direct solar heat transfer through glazing surfaces that is the product of glazing surface, beam component of the total irradiance, solar heat gain value of the glazing system, and level of internal shade provided for the glazing system. Sum of diffused and reflected components of the total irradiance will be multiplied with the glazing system area, diffused solar heat gain coefficient of the glazing area, and the level of internal shading provided for the glazing system to develop the diffused heat transfer through the glazing system (Of course a simple multiplication of glazing area, glazing system U-value, and the temperature difference between the outdoor and indoor environments will be used to calculate the conduction heat transfer through the glazing system as well.).

A detailed calculation of these radiation components is provided in ASHRAE Fundamental Handbook [6] for further study and use.

Convection

"Convection, sometimes identified as a separate mode of heat transfer, relates to the transfer of heat from a bounding surface to a fluid in motion, or to the heat transfer across a flow plane within the interior of the flowing fluid" [4]. DOE Fundamentals Handbook [5] defines convection as transfer of heat between a surface at a given temperature and fluid at a bulk temperature. For flow adjacent to a hot or cold surface bulk temperature of the fluid is far from surface, for boiling or condensation bulk temperature is the saturation temperature, and for flow in pipe bulk temperature is the average temperature measured at a particular cross section of the pipe.

As it was noted earlier, Newton's law of cooling defines the convective heat transfer rate from a surface with an area of A and in contact with a fluid in motion as:

$$q = h \times A \times (t_s - t_\infty) = (t_s - t_\infty)/(1/h \times A) \qquad (1.21)$$

Where h is the heat transfer coefficient (Btu/h ft^2 °F), $1/(h \times A)$ is the convection resistance (h °F/Btu), and t_∞ is fluid temperature and t_s is surface temperature (°F). Since heat always travels from hot region towards the cold region, the temperature difference portion of the equation should be written in such a manner to allocate a positive sign for heat transfer from hot to cold media. Convective heat transfer due to the external forces is known as forced convection and convective heat transfer as the result of the buoyant forces between hot and cold regions is known as natural convection. Therefore, it can be seen that the natural convection usually happens when a fluid exposes to a surface that has a different temperature from the fluid temperature due to its density gradients. Since the natural convection heat transfer coefficient of gases are considerably lower than their forced convection heat transfer coefficient, it has been recommended to include radiation heat transfer which is typically in close approximation of the magnitude of natural heat transfer quantity to be included in natural convection calculations as well. To the contrary and when we are dealing with forced convection as the result of operation of a pump or fan, since the magnitude of heat transfer via forced convection is much larger than the radiative heat transfer quantity, the radiative part of heat transfer can usually be ignored [1].

Convection is one of the most complicated areas of research, and even though some of the equations used in solving convection problems have been mathematically derived and proven, but most of the equations which are commonly used to solve convection problems are purely experimental. Therefore, as a user we can search for the proper model that matches our problem in hand and then look for the proper existing equations relevant to the specific model and solve the problem with a relatively little effort. As it can be seen from Newton's law of cooling the main target of any convection problem solving is to define the proper heat transfer coefficient (h_c). Different dimensionless numbers such as Nusselt, Prandtl, Reynolds, and Rayleigh are developed that combine the multi-characteristics of convective heat transfer, and fit it in a set of functions similar to the ones represented below for a typical forced convection and natural convection. We can calculate Reynolds, Prandtl, and Rayleigh numbers, then use the related function to calculate Nusselt number (Nusselt number has a direct relation with the heat transfer coefficient as it is shown in Eq. 1.22 below) and then from there we can calculate convective heat transfer coefficient.

A sample of equation for forced convection can be represented similar to the one here:

$$\mathrm{Nu} = h \times L_c / k = f(\mathrm{Re}_{Lc}, \ \mathrm{Pr}) \tag{1.22}$$

Where:
Nu = Nusselt number
h = convection heat transfer coefficient
L_c = length characteristic
$\mathrm{Re}_{Lc} = \rho V L_c / \mu = V L_c / \nu$
V = fluid velocity
Pr = Prandtl number = $C_p \times \mu / k$
C_p = fluid specific heat
μ = fluid dynamic viscosity
ρ = fluid density
ν = kinematic viscosity = μ / ρ
k = fluid conductivity

A sample of equation for natural convection can be represented similar to the one here:

$$Nu = h \times Lc \,/k = f(Ra_{Lc}, Pr) \tag{1.23}$$

where

Nu = Nusselt number

h = convection heat transfer coefficient

L_c = length characteristic

K = fluid thermal conductivity

Ra_{Lc} = Rayleigh number = $g \times \beta \times \Delta t \times L_c^3 / \nu \times \alpha$

$\Delta t = |t_s - t_\infty|$

g = gravitational acceleration

β = coefficient of thermal expansion

ν = fluid kinematic viscosity = μ/ρ

α = fluid thermal diffusivity = $k/\rho \times C_p$

Pr = Prandtl number = ν/α

Forced-air coolers and heaters, forced-air- or water-cooled condensers and evaporators, and liquid suction heat exchangers are examples of equipment that transfer heat primarily by forced convection. Some common forms of natural convection heat transfer can be observed in baseboard space heating and cooling panel applications [1].

Based on the location of the fluid inside a pipe, channel, or duct, or outside and over a surface the forced convection is divided to two types: internal and external flow. Reynolds number is the dimensionless number that represents if the flow is laminar or turbulent. For internal flow, laminar boundary layers on each interior wall of the duct or all around the interior surface of the pipe develop and grow until they reach together. Generally for Reynolds numbers (Re = $V_{avg} D/\nu$) calculated based on the diameter for round pipes (or hydraulic diameter for rectangular ducts (Dh = 4 × Cross-sectional area for flow/Total wetted perimeter)) below 2400 the flow is called laminar (flow after reaching together is called fully developed laminar flow) and if flow inside the boundary layer before reaching each other reaches to a Reynolds number of about 10,000, the flow is called turbulent.

For an external flow over a plate the boundary layer generally starts near the plate and grows to turbulent boundary layer near a Reynolds number (Re = $V \times c/\nu$) of 500,000. Heat transfer starts to increase quickly as the transition from laminar to turbulent boundary layer starts.

References

1. Holman, J. P. (2010). *Heat transfer* (10th ed.). New York, NY: McGraw-Hill.
2. Bergman, T. L., Lavine, A. S., Incropera, F. P., & Dewitt, D. P. (2011). *Fundamentals of heat and mass transfer*. New York, NY: Wiley.
3. Lienhard, J. H., & Linehard, J. H. (2008). *A heat transfer text book* (3rd ed.). Cambridge, MA: Phlogiston Press.
4. Rohsenow, W. M., Hartnett, J. P., & Cho, Y. I. (1998). *Handbook of heat transfer*. New York, NY: McGraw-Hill.
5. U.S. Department of Energy. (1992). *DOE Fundamentals Handbook, Thermodynamics, Heat transfer, and fluid flow* (Vol. 2 of 3). Washington, DC: U.S. Department of Energy.
6. American Society of heating, Refrigerating and Air-Conditioning Engineers (2013) ASHRAE handbook fundamental. American Society of Heating, Refrigerating and Air-Conditioning Engineers, Atlanta, GA

Chapter 2
Load Calculations and Energy Modeling

Abstract In designing the HVAC system for a building the first step is always to calculate the heat gain and heat loss during cooling and heating periods. From that point on, the designer will be capable of understanding and deciding what type and what size equipment should be selected to fulfill the comfort and functional criteria in the building. ASHRAE Fundamentals Handbook has been proposing and updating its recommended approaches for load calculation during the past few decades.

Keywords Total equivalent temperature differential method • Sol-air temperature • Transfer function method • Conduction transfer function • Time averaging method • Room transfer function • Cooling load factor • Cooling load temperature difference • Heat balance method • Radiant time series • Base building • Design building

Load Calculations

In designing the HVAC system for a building the first step is always to calculate the heat gain and heat loss during cooling and heating periods. From that point on, the designer will be capable of understanding and deciding what type and what size equipment should be selected to fulfill the comfort and functional criteria in the building. ASHRAE Fundamentals Handbook has been proposing and updating its recommended approaches for load calculation during the past few decades. In the early 1970s ASHRAE proposed total equivalent temperature differential method (TETD) in which data for some of the general envelope assemblies is developed and used. Data is derived from these elements in order to be used to calculate the total equivalent temperature differential as the function of sol-air temperature. Sol-air is defined as the external surface temperature of the envelope as the result of the outside temperature, solar radiation, surface roughness, etc. Total equivalent temperature difference is used to calculate contribution of different elements of the envelope in the total heat gain. Internal load then will be added to this heat gain quantity to represent the instantaneous space heat gain. In order to convert this instantaneous heat gain to instantaneous cooling load the radiation portion of the heat gain is averaged within the current hour and a few hours before that by using a

© Springer International Publishing Switzerland 2016 15
J. Khazaii, *Advanced Decision Making for HVAC Engineers*,
DOI 10.1007/978-3-319-33328-1_2

time averaging (TA) method. Furthermore and in the late 1970s ASHRAE represented another method that is called transfer function method (TFM) in which conduction transfer function (CTF) is used as a weighting factor in order to include the proper thermal inertia of the opaque elements of the envelope into the calculation mix, and room transfer function (RTF) is used as a weighting factor in order to include the proper thermal storage effect in converting heat gain to cooling load. In this method solar heat gain and internal loads are calculated instantly. In the late 1980s ASHRAE proposed utilizing a new method which is called cooling load factor (CLF) by defining cooling load temperature difference (CLTD) for walls and roofs, solar cooling load (SCL) for glasses, and cooling load factor (CLF) for internal load elements. Therefore, in this method which is a simplified version of the earlier methods, the important factors are the time lag in conduction heat gain through the exterior walls and roofs, and time delay due to thermal storage where the radiant heat gain is converting to cooling load. These factors are included in calculations with a simpler approach which is basically utilization of predeveloped tables. For heat gain through walls and roofs, both TFM and TATD-TA methods use the concept of sol-air temperature similar to CLTD/SCL/CLF method. TFM method further uses another set of multipliers to convert sensible loads to cooling load based on the storage capacity of the objects, walls, etc. in the space. Meanwhile TATD-TA method uses supportive equation to include the effect of time lag in walls and roofs in calculations, and uses an averaging method for calculating radiation part of the sensible loads. CLTD/SCL/CLF method uses also sol-air concept and most other assumptions made in TATD-TA method to develop appropriate simply usable tables for CLTD, SCL and CLF for some specific walls. In this method, ASHRAE recommends where specific walls or roofs are not listed, the closest table should be used. A detailed load calculation process using all of these methods can be found in ASHRAE Fundamentals Handbook 1997 [1]. Another method which has been around for a while but due to its relatively complicated approach has not been used practically for load calculation is heat balance (HB) method. "Although the heat balance is a more scientific and rigorous computation, there are limitations that result from the high degree of uncertainty for several input factors. In many cases, the added detail required of this method is not warranted given the variability and uncertainty of several significant heat gain components" [2].

In 2001, ASHRAE Fundamentals Handbook [3] used the heat balance method structure as the basis for developing another useful and less complicated method which is called radiant time series (RTS). Here I attempt to shed some light on procedure of implementation of this method and briefly explain its approach to load calculations since it is basically the currently accepted method by ASHRAE Fundamentals Handbook for load calculation.

In this method the main deviation from the original heat balance method (HB) and of course in order to decrease the intensity of the calculations is the assumption that the 24 h condition pattern repeats itself throughout the month and therefore heat gain for a particular hour is the same for the previous days at the same particular hour.

In this method the first step similar to the previously used cooling load calculation methods is to calculate the solar intensity on all exterior surfaces in every hour, and use this solar intensity to calculate the associated sol-air temperatures. This sol-air temperature in conjunction with an appropriate conduction time series is used to calculate heat gain for each hour at each exterior surface. Transmitted, diffused and ground reflected solar heat gains and conductive heat gain for the glazing, and also heat gain through lighting, equipment and people should be calculated as well. In the next step all the radiant and convective heat gains should be separated. Total convective portion should be added to the processed radiant portions which is calculated based on appropriate radiant time series to represent the final hourly cooling load.

ASHRAE Fundamentals Handbook [3] defines the sol-air temperature as combination of the outdoor air temperature (°F) with two other factors. The first factor is an addition to the outdoor air temperature and is defined as the multiplication of absorptance of surface for solar radiation (α-dimensionless) by total solar radiation incident on surface in (Btu/h ft^2) divided by coefficient of heat transfer by long-wave radiation and convection at outer surface in (Btu/h ft^2 °F), and the second factor is a deduction from the outdoor air temperature and is defined as multiplication of hemisphere emittance of surface (ε) by difference between long-wave radiation incident on surface from sky and surrounding and radiation emitted by blackbody at outer air temperature in (Btu/h ft^2) divided by coefficient of heat transfer by long-wave radiation and convection at outer surface in (Btu/h ft^2 °F). By this definition one can observe that ASHRAE by presenting sol-air temperature is providing a simple substitute temperature which in absence of all radiation changes provides the combination effect of heat entry into the surface which is similar to the combined effect of incident solar radiation, radiant energy exchange with the sky and outdoor surroundings, and convective heat exchange with outdoor air.

ASHRAE Fundamentals Handbook [3] recommendation for the deductive portion of the equation representing sol-air temperature is to approximate this value as 7 °F for horizontal and 0 °F for vertical surfaces since in most cases this generalization creates a very good approximation for sol-air temperature without extensive calculations, and makes the calculations for sol-air temperature much simpler. The multiplier to the addition portion of sol-air temperature also can be estimated to constant values based on the lightness or darkness of the wall material. This sol-air temperature then will be used as the overall external surface temperature of walls and roofs. A simple equation of conduction then can be executed which is the multiplication of the area of the exterior wall or roof by its associated U-value and by the temperature difference between calculated sol-air temperature and interior space surface temperature to represent the whole heat transfer through the wall and roof.

RTS method suggests to calculate this heat transfer for all the 24 h of the day and use the given time series multipliers to separate actual portion of the heat to be considered for each hour due to time lag associated with the conduction process. Therefore, after doing this action at each instance we can have the real heat that is

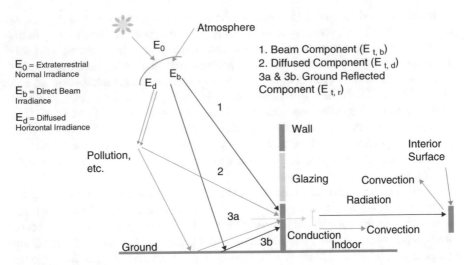

Fig. 2.1 Sun effect through opaque surface

transferred to the interior space by summing the associated portion of the heat transfer at that instance and all the past 23 h instances before that.

The idea is that one part of this transferred heat will be immediately transferred to the room air by convection and becomes instant cooling load, while the other portion of it will be in the form of radiation that requires to be radiated on and be absorbed by the interior walls and objects and then be transferred back to the space air through convection as well. The heat gain share of this process also becomes instant cooling load in the space. This concludes the cooling load calculation due to the exterior walls and roofs (Fig. 2.1).

ASHRAE Fundamentals Handbook [3] RTS method suggested approach for calculating instant cooling load through glazing of the building includes calculating surface beam irradiation, diffuse irradiation and ground reflected irradiation by using beam normal irradiation and the appropriate equations in addition to the simple heat gain calculation via conduction through the glass. If there is no interior cover for the glazing such as window cover, then all the heat gain through surface irradiation becomes radiant type, and therefore, the entire heat gain through this will be absorbed by the interior areas and then heat will be transferred to the space air via convection as instant cooling load. Surface irradiation along with glazing area and associated solar heat gain coefficient should be used to calculate hourly heat gain and then radiant time series should be implemented in calculations to provide the proper rate of this transfer at each hour due to the radiation time delay.

Similar to surface beam irradiance, diffuse and ground reflected irradiance along with glazing area and associated solar heat gain coefficient should be utilized to calculate the second portion of heat gain through the glazing. Only exception is that this time we should separate the convection and radiant portion of the heat gain via recommended percentages provided by ASHRAE Fundamentals Handbook [3]. Convection portion will become immediate cooling load and the radiant portion should be

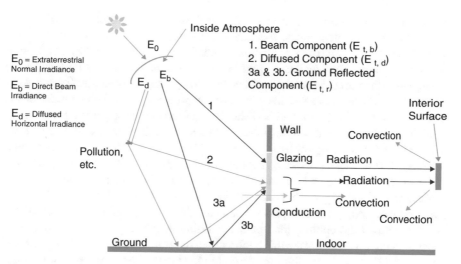

Fig. 2.2 Sun effect through glazing

adjusted via radiant time series to allow for the proper heat gain due to time delay in heat gain change to instant cooling load. It should be noted that if interior covers such as window blinds are provided, the surface beam irradiance should be treated similar to diffuse and ground reflected irradiance. Instant cooling calculated through these processes should be added to the simple conduction heat transfer through glazing (multiplication of glazing area by U-value of the glazing by the temperature difference between outdoor and indoor) to provide the total instant cooling to the space through glazing system (Fig. 2.2).

To conclude calculation of instant cooling load in space, RTS method recommends using time series for calculating instant cooling from people, lighting and equipment sensible loads. Proper multipliers for dividing the convective and radiant portions of these sensible loads are provided by ASHRAE Fundamentals Handbook [3], and similar to what has been described earlier, the convective portion (along with latent heats from people and equipment) becomes instant cooling load, and the radiant part should be adjusted via using proper time series to define the final portion of the space instant cooling load.

In order to clarify the method a little more, let us now further look at the instantaneous cooling load calculation by using the RTS method with a step by step procedural explanations. When we are dealing with the internal loads, other than the latent heat load which becomes instantly part of the cooling load, in calculating the instant cooling load, contribution of the sensible cooling load of the internal loads (lighting, equipment, and people) the following procedure should be followed. At first we should develop the 24 h heat gain profile of the internal load (e.g., lighting) based on the respected operating schedule for lighting and equipment or occupancy for people. Assume the calculated lighting load for the targeted hour is (880 W × 3.4 (btu/h)/W × 100 % usage) 3000 btu/h. This load should be separated to radiation and convection portions. This multiplier is given by ASHRAE in associated tables. Assume a 48 % share of the radiation part, and it will result in (3000×0.48) 1440 btu/h radiation and (3000×0.52) 1560 btu/h convective share. The convective part becomes instantaneous cooling load to the space, while the radiation share should be treated by proper time series for non-solar RTS values. These values are given in tables as percentages of the radiation for the targeted hour and its previous 23 h. It should be noted that these percentages sum up to a 100 % for the current and the previous 23 h (e.g., multipliers are: at current hour or hour 0, 49 %; at 1 h ago or hour 1, 17 %; at 2 h ago or hour 2, 9 %; . . .; at 23 h ago or hour 23, 0 %). Using these multipliers in accordance with the associated 24 h radiant part of the load and summing those up will produce the radiation part of the cooling load at the targeted hour. By adding the cooling load contribution from the radiation part and convective part we can calculate the overall contribution of the lighting load to the cooling load of the building at the targeted hour. Of course depending on how we allocate this cooling load to return air and inside room air, we can calculate the final contribution of lighting load to the instant room cooling load. Similar procedures should be followed to calculate the other internal loads contributions to the building overall instantaneous cooling load.

The next step naturally is to calculate the contribution of each opaque surface to the instantaneous cooling load to the building using procedure presented by RTS method. The first step is then to calculate sol-air temperature for each of the 24 h in the targeted month on each surface. In each hour then we can calculate the quantity of the conducted air through the targeted surface by multiplying the area and U-value of the surface by the difference between the sol-air temperature of the exterior surface and the interior surface temperature which is usually equal to the room air temperature. Such exercise produces a set of 24 h of the conducted heat from exterior to interior surface. Then we have to use a conductive time series (CTS) multipliers which represents the portion of contribution of current targeted hour and the previous 23 h for the conduction through the wall at the targeted (current) hour (e.g., multipliers are: at current hour or hour 0, 18 %; at 1 h ago or hour 1, 58 %; at 2 h ago or hour 2, 20 %; . . .; at 23 h ago or hour 23, 0 %). Assume we have calculated 700 btu/h heat gain for current hour, 600 btu/h for previous hour, etc. Then the total conductive load through the wall for the targeted hour is the sum of the results of the multiplication of the multipliers by their associated calculated heat gain in the past 24 h (e.g., $700 \times 0.18 + 600 \times 0.58 + \ldots + 750 \times 0$). Once again assume the result of this summation for the current hour is 650 btu/r (i.e. 18 % of the current hour conduction, plus 58 % of the previous hour conduction, . . ., plus 0 % of the 23 h ago conduction is equal to 650 btu/h). This conducted sum for this hour should then be separated to convection and radiation portions according to the percentages given in ASHRAE tables. Assume the tables assign a 54 and 46 % shares to convection and radiation portions, and we will have (0.54×650) or 351 btu/h for convection part and (0.46×650) or 299 btu/h for radiation part. Convection part becomes instant cooling load for the building but the radiation part should be treated by RTS multipliers before is considered as instant cooling load. Such calculations should be repeated for all the other 23 h of the month.

Now let us divert our attention towards glazing system contribution to the instant cooling load as it is defined by RTS method. ASHRAE Fundamental Handbook [3] has presented a set of equations that can be utilized in order to calculate direct solar beam, diffused beam and conductive heat gain through glazing systems. We have also briefly discussed these methods in earlier chapter of this book. ASHRAE recommends to put these equations in use to calculate the 24 h profile of direct solar beam, diffused beam and conductive heat gain through the glazing on each surface of the building at its designated orientation. The proposed method then is to use the direct solar beam (assuming there is no internal shading) as 100 % radiation type. This calculated direct solar beam should be treated by RTS solar factors to calculate the share of the current hour and the previous 23 h of direct solar beam on the instantaneous cooling load from the direct solar beam. These factors define what percent of the calculated heat gain for this hour and the past 23 h should be summed up to generate the instant cooling for current hour (e.g., multipliers are: at current hour or hour 0, 54 %; at 1 h ago or hour 1, 16 %; at 2 h ago or hour 2, 8 %; . . .; at 23 h ago or hour 23, 0 %). Using these multipliers in accordance with the associated 24 h radiant heat gain at each hour and summing those up will produce the contribution of the direct solar beam on the cooling load at the targeted hour. Assume we have calculated 3000 btu/h heat gain for current hour, 2000 btu/h for previous hour, 1800 btu/h for 2 h ago, etc. Then the total radiant load through the glazing for the targeted hour is the sum of the results of the multiplication of the multipliers by their associated calculated heat gain in the past 24 h (e.g., $3000 \times 0.54 + 2000 \times 0.16 + 1800 \times 0.08 + \ldots$).

Next we have to sum up the shares of diffused solar beam and conductive heat gain through the glazing system for the past 24 h ending in current targeted hour. This heat gain then should be separated to radiation and convective portion via presented percentages offered by ASHRAE Fundamental Handbook [3] tables. The convective portion will become immediate cooling load for the space, while the radiation portion should be treated by RTS non-solar time series to calculate the current hour instant cooling considering the share of this value based on the past 24 h contributions. The sum of the direct solar beam, diffused beam and convective cooling loads will represent the contribution of the glazing on instant cooling load of the space. It should be noted that if there is inside shading provided for the glazing system, direct solar beam effect should be treated similar to diffused solar beam.

ASHRAE Fundamentals Handbook [3] recommends this method as the replacement for all the previous cooling load methods for calculating the maximum cooling load and selection of the equipment, but does not recommend this method for energy modeling calculations due to its restrictive and simplifying assumptions. Therefore, RTS method should be used for load calculation only while any one of the previously mentioned cooling load methods could be used for energy modeling.

Of course the heating load calculation is much simpler than cooling load calculation. It can be simplified and summarized to as trivial as multiplying exterior exposure surfaces by their associated U values by the temperature difference between the outdoor and the inside space.

A detailed procedure for load calculation by RTS method has been provided in ASHRAE Fundamentals Handbook 2001 [3] and later versions.

Energy Modeling

Building energy modeling is the basic procedure that could help predicting how the building as a whole—including its associated systems and components—from the standpoint of energy usage may operate after it is constructed. Higher energy consumption of the building during operation in addition to resulting in more environmental undesired consequences, has direct cost increase implications for the owner of the building as well. For these reasons tremendous attention has been focused towards developing guidelines, procedures and software to accommodate performing energy modeling that can predict the operation of the building systems more accurately and to help the architects and engineers to select more energy efficient strategies, components, and systems when designing a building.

In order to be able to run a reliable energy model for a building the modeler needs (1) a very good understanding of the building and its associated systems, (2) an in-depth understanding about the guidelines that regulates modeling procedure, and of course (3) a very good understanding of how to use the software that is used for the modeling properly. Lack of understanding of either of these items surely will lead to unreliable results from the models. Therefore, it is crucial for the reliability of the results that only the qualified persons in each firm to be selected to develop the energy models.

After acknowledging this, let us start our describing energy modeling procedure with a simple multi-story (e.g., six story) office building, which is one of the most straightforward and simplest building types from the stand point of commercial buildings design. Typical office buildings usually include multiple open space offices, single executive offices, conference rooms, front lobby and corridors, copy rooms, computer rooms, etc. Traditionally a core section is assigned that includes restrooms, elevators, mechanical rooms, electrical closets, audio visual closets and general vertical open shafts for running ducts up and down the building. Usually a fire rated enclosure separates the core area from the so called shell area where the actual offices and other building specific areas are located at. The main purpose of this rated enclosure is to separate multi-story related core from the shell in case of breaking fire in the building. At each duct penetration through the rated enclosure proper fire dampers should be installed to keep the integrity of the rated portion all around.

Architect combines the owner's desire with the architectural regulations to lay-out the office building with a mixture of these elements. The architectural set of drawing which are delivered to the HVAC engineer (along with other engineers such as electrical and Plumbing engineers) show the elevations, sections, floor plans, furniture plans, ceiling plans, site plans, wall and roof construction detail, etc. These information along with the guidelines from ASHRAE Standard 90.1 [4] will be used to set up the energy model for the building. Based on the guidelines of this standard, each building can be separated to thermal zones and even though sometimes a detailed room by room energy model may be very helpful but in most conditions it is not necessary to provide a room by room detail model for modeling calculations. Based on this guideline, the design building energy model shall be compared against the energy model of a base-building which is very similar to the design building shape but it has some differences in regards to the required systems and some component. This base building model tends to represent a threshold to be surpassed by the design building in energy conservation measures. Of course the higher the efficiency of the design building (lower energy consumption and cost) is compared to the efficiency of the base-building the better is the performance and efficiency of the design building based on guidelines of this standard.

Two of the most popular software for running energy modeling in professional world are e-Quest and Trane Trace 700. The e-Quest software uses a three floor model for multiple story building energy modeling and represents a graphical view of the building as well. In e-Quest the first floor and top floor are modeled separately and all other middle floors are usually modeled together as one big floor. If a building is simple rectangular shape as we considered for our example, then the whole building can be separated into a total of 15 thermal zones. Four exterior and one interior zones are assigned to the first floor, another four exterior and one interior thermal zones are assigned to all the middle floors and four exterior and one interior (this final so-called interior zone practically has exterior exposure through the its roof, and therefore, it is not a complete interior zone) thermal zones will be assigned to the top floor. We can separate the building into thermal zones similarly in Trane Trace 700 as well. The only difference is that middle floors in Trane Trace 700 are usually assigned separately.

Due to the degree of importance of effects of building glazing in the exterior 15 ft of each floor, it is customary to allocate a 15 ft depth to each exterior thermal zone (as long as there is no separating wall within this 15 ft depth) and consider the rest of the floor space as internal zones. It is important not to mix the terminal units serving zones with different exposures, but multiple terminal units can discharge air to one single thermal zone depending on the size of the zone. Therefore, it is customary to assign two or three terminal units to each thermal zone during building design process.

Different HVAC systems can be selected to provide air conditioning for an office building. To name a few are variable volume air handling system with chilled and hot water producing plants deploying chillers and boilers, self-contained direct expansion variable volume air handling systems with condenser water heat rejection system and hot water coil or electric strips deploying cooling towers and boilers if needed, and simple packaged direct expansion variable volume air delivery system with hot water or electric heating strips. As it stands currently variable air volume system is the basic dominating system for air delivery to such buildings since the constant air volume air handling units are not energy efficient for such purposes. Other air delivery systems such as under-floor systems can be used in office buildings as well. Even though the technique has been frequently used

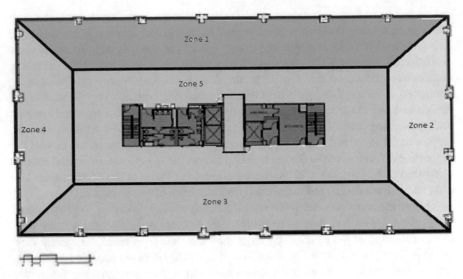

Fig. 2.3 Typical building thermal zoning

for air conditioning computer rooms, but it has not widely opened its way to the general office building applications yet. Similarly passive or active chill beam systems are also new technologies that have not been used widely in practice world due to the current market structure. Other major parts of the system or cooling and heating plants still should be selected as well.

When the designer is not forced to design any specific system due to the source availability, delivery time restrictions, agreements or space limitations, it is proper to run energy modeling in order to justify selecting a specific system over other alternatives.

To start our energy modeling experiment assume we have the six story office building in Atlanta (Georgia) for our study. And also assume each of the building six floors has a typical shell floor plan similar to the one shown in Fig. 2.3.

The approach that I have taken here is more appropriate for a preliminary energy modeling in early stages of the design where the detail of the space lay-out is not available, but generally and even after a detailed lay-out of the building is available, due to thermal zone approach instead of room by room modeling, this approach provides very good approximation of energy consumption of the building.

As we can see in the core space of each floor a space (mechanical room) has been assigned as mechanical space that can be utilized as the home to an air handling unit designated to serve that specific floor. As I noted earlier, it is a customary design to include the emergency stairs, elevators shafts and lobby, public restrooms, electrical and audio-visual closets in this central space in each floor as well. In this building there is no available space currently designated to the heating and cooling plants inside the foot print of the building, and therefore, space next to the building or above the roof has to be designated for this purpose, depending on what type of heating or cooling plant is selected for the design project.

The followings are the type of information and some possible value (these values vary based on the engineer's choice of materials and systems) associated with these information for our multistory building in Atlanta to be used in a typical (Trane Trace 700) energy modeling software. This model is usually called design or proposed building. The values used here are the values that the designer specifies for the building to be designed with. Later on we will show the changes required to these information in order to revise it to what in ASHRAE 90.1 Standard [4] is called base building. The target is to specify by what percent the proposed building energy consumption (cost) can beat the base building energy consumption (cost).

Typical input data to the proposed building energy model would be as follows.

Interior Space Conditions

Cooling season: 75 F dry bulb and 50 % relative humidity.
Heating season: 72 F.
Floor to floor height: 13 ft.
Wall height: 13 ft.
Plenum depth: 4 ft.
Exterior wall U-value: 0.08 Btu/h ft^2 °F
Roof U-value: 0.04 Btu/h ft^2 °F
Glass U-value and Solar Heat Gain Coefficient: 0.65 Btu/h ft^2 °F, 0.29 (60 % window to wall ratio on all sides)
Slab on grade: 0.73 Btu/h ft^2 °F
Partitions: 0.4 Btu/h ft^2 °F

General office internal loads

Number of people: General office space (200 sq ft/person, 250 Sensible Btu/h, 200 Latent Btu/h)
People Schedule: People—Mid-rise Office
Lights: Recessed Fluorescent, not vented, 80 % load to space (1.0 W/ sq ft)
Light Schedule: Light—Mid-rise Office
Miscellaneous Load: 0.5 W/ sq ft
Miscellaneous Load Schedule: Miscellaneous Mid-rise Office

General office Ventilation

Ventilation: Based on guidance of ASHRAE Standard 62.1 [5] (General office spaces 5 cfm/person plus 0.06 cfm/ft^2)
Ventilation type: Office Space; Cooling EZ: Ceiling Supply 100 %, Ceiling Return; Heating Ez: Ceiling Supply < room temperature plus 15 °F, 100 %.
Ventilation Schedule: Ventilation Mid-rise Office
VAV Minimum Stop: 30 %
Infiltration: 0.15 Air changes/h
Infiltration Schedule: Infiltration Mid-rise Office

Conference rooms internal loads

Number of people: (20 sq ft/ person, 250 Sensible Btu/h, 200 Latent Btu/h)
People Schedule: People—Mid-rise Office
Lights: Recessed Fluorescent, not vented, 80 % load to space (1.0 W/ sq ft)
Light Schedule: Light—Mid-rise Office
Miscellaneous Load: 0.5 W/ sq ft
Miscellaneous Load Schedule: Miscellaneous Mid-rise Office

Conference rooms ventilation

Ventilation: Based on guidance of ASHRAE Standard 62 (Conference rooms
    5 cfm/person and 0.06 cfm/ft^2)
Ventilation type: Conference rooms; Cooling EZ: Ceiling Supply 100 %, Ceiling
    Return; Heating Ez: Ceiling Supply < room temperature plus 15 °F, 100 %.
Ventilation Schedule: Ventilation Mid-rise Office
VAV Minimum Stop: 30 %
Infiltration: 0.15 Air changes/h
Infiltration Schedule: Infiltration Mid-rise Office

Lobby internal loads

Number of people: General office space (100 sq ft/ person, 250 Sensible Btu/h,
    200 Latent Btu/h)
People Schedule: People—Mid-rise Office
Lights: Recessed Fluorescent, not vented, 80 % load to space (1.0 W/ sq ft)
Light Schedule: Light—Mid-rise Office
Miscellaneous Load: 0.5 W/ sq ft
Miscellaneous Load Schedule: Miscellaneous Mid-rise Office

Lobby ventilation

Ventilation: Based on guidance of ASHRAE Standard 62 (General office spaces
    5 cfm/person and 0.06 cfm/ft^2)
Ventilation type: Lobby; Cooling EZ: Ceiling Supply 100 %, Ceiling Return;
    Heating Ez: Ceiling Supply < room temperature plus 15 °F, 100 %.
Ventilation Schedule: Ventilation Mid-rise Office
VAV Minimum Stop: 30 %
Infiltration: 0.15 Air changes/h
Infiltration Schedule: Infiltration Mid-rise Office

Corridors and Restrooms internal loads

Number of people: General office space (0)
People Schedule: People—Mid-rise Office
Lights: Recessed Fluorescent, not vented, 80 % load to space (1.0 W/ sq ft)
Light Schedule: Light—Mid-rise Office
Miscellaneous Load: None
Miscellaneous Load Schedule: None

Corridors and restrooms ventilation

Ventilation: Based on guidance of ASHRAE Standard 62 (General office spaces
0 cfm/person and 0.06 cfm/ft^2)
Ventilation type: Corridor; Cooling EZ: Ceiling Supply 100%, Ceiling Return;
Heating Ez: Ceiling Supply < room temperature plus 15 °F, 100%.
Ventilation Schedule: Ventilation Mid-rise Office
VAV Minimum Stop: 30%

Room (thermal block) assignments

Spatial information for each block of similar function spaces (offices, conference
rooms, etc.) in each thermal zone should be developed and the characteristics
explained above should be designated to them. These thermal blocks then should
be assigned to the proper air system as well.

Airside System: (This should be repeated and revised for each actual building
system.)

System Type: Variable Volume Reheat
Supply Air Filtration: MERV 13 filter
Return Air Path: Plenum return
Economizer Type: Enthalpy type
Optimum Start: Available (100%)
Supply Air Temperature Reset: Available (100%)
Cooling Supply Air Temperature: To be calculated
Primary Supply Fan: FC Centrifugal with VFD
Primary Fan Energy: 30 hp (92.4% motor efficiency)
Supply Fan Static Pressure: 3.5 in. of water gauge
Exhaust Fan: FC Centrifugal with VFD
Exhaust Fan Energy: 20 hp (92.4% motor efficiency)
Exhaust Fan Static Pressure: 1 in. of water gauge
Fan Cycling Schedule: Cycle with Occupancy

Cooling Plant

Equipment Type: 480 t—Water Cooled Unitary, Cooling tower with VFD and
1440 gpm variable volume condenser water pumps.
Sequencing Type: Single
Efficiency: 0.797 kW/t

Heating Plant

Equipment Type: Electric resistance strips
Capacity: 500 kW, efficiency 98%

Miscellaneous Utility loads

Elevator Machine room, Gas fired water heater load, etc. (inputs as required)

Utility Rates

Electricity (consumption): $100 base charge plus $0.1/kWh
Electricity (Demand): $10/kW
Gas (rate): $100 base charge plus $0.5/therm
Water (rate): $ 5/k-gallon

As it was noted earlier, ASHRAE 90.1 Standard [4] G-section requires a second energy modeling of the building known as "Base Building" model in order to compare its energy consumption (coat) results against the proposed building results and verify if the proposed building is meeting the requirements of energy standard, and if it is meeting this standard by how much it is passing the minimum required threshold. Typical revised input data to the base building energy model compared to the proposed building input data is as follows:

Interior Space Conditions

Exterior Wall U-value: 0.077 Btu/ h ft^2 °F (based on ASHRAE 90.1 Standard building envelope requirement for specific climate zone)
Roof U-value: 0.039 Btu/h ft^2 °F (based on ASHRAE 90.1 Standard building envelope requirement for specific climate zone)
Glass U-value & Solar Heat Gain Coefficient: 0.5 Btu/h ft^2 °F, 0.25 (40 % window to wall ratio on all sides) (based on ASHRAE 90.1 Standard building envelope requirement for specific climate zone)
Slab on grade: 0.73 Btu/h ft^2 °F (based on ASHRAE 90.1 Standard building envelope requirement for specific climate zone) (see Table 2.1 below).

General office internal loads

Number of people: General office space (200 sq ft/person, 250 Sensible Btu/h, 200 Latent Btu/h) (based on ASHRAE 62.1 Standard building occupancy requirement for specific (office) application)
People Schedule: People—Mid-rise Office

Table 2.1 Glazing distribution, proposed vs. base building

	Proposed		Baseline		Ratio
Gross exterior wall	48,000	ft^2	48,000	ft^2	
Window wall ratio	60	%	40	%	
Window area	28,800	ft^2	19,200	ft^2	0.67
North wall	9600	ft^2	6400	ft^2	
South wall	9600	ft^2	6400	ft^2	
East wall	4800	ft^2	3200	ft^2	
West wall	4800	ft^2	3200	ft^2	
Net wall	19,200	ft^2	28,800	ft^2	

Lights: Recessed Fluorescent, not vented, 80 % load to space (0.87 W/sq ft) (based on ASHRAE 90.1 Standard building maximum lighting allowance for specific (office) application)
Light Schedule: Light—Mid-rise Office

Conference rooms internal loads

Number of people: (20 sq ft/ person, 250 Sensible Btu/h, 200 Latent Btu/h) (based on ASHRAE 62.1 Standard building occupancy requirement for specific application)
People Schedule: People—Mid-rise Office
Lights: Recessed Fluorescent, not vented, 80 % load to space (1.23 W/sq ft) (based on ASHRAE 90.1 Standard building maximum lighting allowance for specific application)
Light Schedule: Light—Mid-rise Office
For Lobby, restroom and corridors follow a similar approach as well.

Airside System

System Type: Variable Volume with PFP (Parallel Fan Powered) boxes) (based on ASHRAE 90.1 Standard base building HVAC system types when a specific system is utilized for proposed building)
Primary Fan Energy: to be calculated based on supply air to room temperature difference of 20 °F. Using the following equation:

$$P_{fan} = bhp \times 746/ \text{ fan motor efficiency} \qquad (2.1)$$

Where:

$$bhp = CFMS \times 0.0013 + \text{pressure drop adjustments} \qquad (2.2)$$

Exhaust Fan Energy: to be calculated similar to the process described above.

Cooling Plant

Equipment Type: Two equally sized water-cooled screw chillers (due to the total cooling capacity larger than 300 t and smaller than 600 t) (capacity to be calculated and multiplied by 1.15 to allows for additional 15 % cooling in base building) (based on ASHRAE 90.1 Standard base building cooling system requirement).
Chilled water pump: 22 W per gpm (based on ASHRAE 90.1 Standard base building cooling system requirement)
Condenser water pumps: 19 W per gpm (based on ASHRAE 90.1 Standard base building cooling system requirement)

Heating Plant

Equipment Type: Electric resistance strips (based on ASHRAE 90.1 Standard base building cooling system requirement)
Capacity: to be calculated

It should be noted that whenever the proposed building data input matches the requirements of the base building data input, the data will remain unchanged. Therefore, the rest of inputs from the proposed building represented above stays unchanged when we set up the base building energy model and it has not been repeated in the base building section presented above for the benefit of time and space. Of course there will be some additional minor changes from the proposed building to the base building model based on the design and the designer's interest, but above I have tried to cover the most important deviations between what an engineer includes in his proposed design and what ASHRAE 90.1 dictates as the requirements of the based building. At the end the energy consumption (cost) of the proposed building will be compared against the energy consumption (cost) of the base building. When sustainability scoring systems such as LEED is targeted the higher the percentage improvement we can get for the proposed building versus the base building, the higher credits will be scored for the building via the sustainability scoring system.

In addition to G-section in ASHRAE standard 90.1 [4] which helps to compare the energy consumption of the simulated design building with a base building, in another section, ASHRAE standard 90.1 [4] section 11 literally follows the same procedure with some minor exceptions. The main difference between section G and section 11 is that the former is used to allow the simulator to score some extra point as he improves the design by increasing the difference between the design and base building energy consumption, but the latter is only used to verify that the energy consumed by the design building is lower than the specified threshold by the standard and therefore allows the design building to be justified as an energy-efficient one.

Other important sections of ASHRAE standard 90.1 [4] to be strictly complied with are the mandatory sections in each of the main sections of envelope, mechanical, plumbing, lighting, electrical and miscellaneous power.

References

1. American Society of Heating, Refrigerating and Air-Conditioning Engineers (1997) ASHRAE handbook fundamental. American Society of Heating, Refrigerating and Air-Conditioning Engineers, Atlanta, GA
2. Kavanaugh SP (2006) HVAC simplified. American Society of Heating, Refrigerating and Air-Conditioning Engineers, Atlanta, GA
3. American Society of Heating, Refrigerating and Air-Conditioning Engineers (2001) ASHRAE handbook fundamental. American Society of Heating, Refrigerating and Air-Conditioning Engineers, Atlanta, GA
4. American Society of Heating, Refrigerating and Air-Conditioning Engineers (2013) ASHRAE Standard 90.1; Energy standard for buildings except low-rise residential buildings. American Society of Heating, Refrigerating and Air-Conditioning Engineers, Atlanta, GA
5. American Society of Heating, Refrigerating and Air-Conditioning Engineers (2013) ASHRAE Standard 62.1; Ventilation for acceptable indoor air quality. American Society of Heating, Refrigerating and Air-Conditioning Engineers, Atlanta, GA

Part II
High Energy Consuming HVAC Applications

Chapter 3
Data Centers

Abstract One of the HVAC most energy-demanding applications is designing a data center. In the current environment and with rapid advancement of technology the need for designing and maintaining data centers can be felt more than ever. A computer room, server room, IT room, or data center is a dedicated space for locating computer systems which handles a large quantity of data such as telecommunications and storage systems.

Keywords Computer room • Server room • IT room • Data center • Telecommunications • Storage systems • Power usage effectiveness

One of the HVAC most energy-demanding applications is designing a data center. In the current environment and with rapid advancement of technology the need for designing and maintaining data centers can be felt more than ever. A computer room, server room, IT room, or data center is a dedicated space for locating computer systems which handles a large quantity of data such as telecommunications and storage systems. The modern and high-density data racks that contain the servers could be powered well above 100 KW [1]. To manage such level of power generation an extremely high capacity cooling system should be designed and installed. "Data center spaces can consume up to 100–200 times as much electricity as standard office spaces. With such large power consumption, they are prime targets for energy-efficient design measures that can save money and reduce electricity use. However, the critical nature of data center loads elevates many design criteria—chiefly reliability and high power density capacity—far above energy efficiency" [1]. In general the current equipment racks which are used in data centers are built as one of the two main types of air-cooled and liquid-cooled units. The capacity of most of the current air-cooled racks are less than 30 KW, while liquid-cooled systems are commonly utilized when the power is considerably higher. The big challenge for a HVAC engineer when designing such spaces is that both the high and low temperatures and humidity limits should be controlled closely, while in most other applications such tight control of all these limits are not likely to be necessary. The reason for such tight control requirement by ASHRAE [2] explained as, allowing the room to operate under high temperature condition will result in a decrease in reliability and also life of the equipment, while maintaining low temperatures obviously may cause extremely more than required

© Springer International Publishing Switzerland 2016
J. Khazaii, *Advanced Decision Making for HVAC Engineers*,
DOI 10.1007/978-3-319-33328-1_3

energy consumption for the system. Furthermore, higher and lower humidity in these spaces could cause failure, error, corrosion, and electrostatic damage to the equipment. Two definitions are usually used by the manufacturers when they attempt to describe the environmental limits of their equipment. Manufacturers use "allowable environment" phrase which is the environment which they test their equipment to make sure their equipment will meet the functional requirements specified by the testing agencies for the testing time period versus the "recommended environment" phrase that is the required environment where the equipment should operate under normal conditions after it is installed and during its life time thereafter. However, sometimes there are some minor allowances by the manufacturer for the equipment to operate outside of the recommended environment range for some limited period of time as well [2].

ASHRAE Application Handbook [5] defines eight group of data center equipment that all demand proper and tightly controlled air conditioning measures, but each with a different level of accuracy separated by different high–low limits of temperature and humidity control. These equipment consists of type 2U and greater, 1U, blade, and custom computer servers, high and extreme density communication, tape storage, storage servers, workstations, and other rack- and cabinet-mounted equipment.

ASHRAE Datacom [2] defines six classes of air-cooled equipment for the data centers. A1 through A4 (different levels of data centers), B (office, home, and transporting environment), and C (points of sale equipment, industrial, factory). Among these categories class A1 is the mission critical class and requires the tightest control of environment, and class C is the lowest class with close to no more than general controls that are similar to a regular office building control requirement. The recommended environment for all four "A" classes are given in this publication, and data center designers and operators have been warned that operating the data center beyond these recommended criteria can cause major damage to the systems, if it lasts long enough.

One of the significant potential drawbacks of air-cooled data center equipment is that the data center equipment usually contain variable volume internal fans that are set by the manufacturer to deliver a specific inlet air temperature to the equipment. This leaves the door open for possible compromise by some design or operational mistakes. Assume a specific situation that the data center air handling unit is intentionally designed by designer for a higher air temperature delivery or when the room temperature set-point has been adjusted up unwisely by the maintenance people during the operation in order to manage a higher than recommended temperature delivery to the unit with the purpose of overall energy saving. Such provisions could result in a complete adverse way such that the rack internal fan operates at a considerably higher speed and therefore consumes much higher level of power than it would if it was operating in an environment with lower and proper temperature specified by the manufacturer.

Since water has higher capacity to absorb heat per volume or mass relative to air in situations that there are real large heat removal requirements due to large data centers power consumption, water loops are much better solutions for cooling the data room equipment [2]. One main problem with this concept is the possibility of the water leakage in presence of electronic powered equipment which can lead to disastrous accidents. Therefore, in situations that we may select water-cooled data center equipment comprehensive preventive approaches should be taken under consideration to eliminate any chance of possible water leakage from the pipes and equipment. In addition to some usually implemented approaches to prevent leakage such as completely enclosing pipes in protective covers or enclosures and enforcing very high level of workmanship, other non-regular approaches such as using thermally, but not electrically conductive liquids in cooling loops instead of water have been adopted as well. ASHRAE Datacom [2] defines five classes (W1–W5) for water-cooled equipment, each with its own associated lower and higher limits of the facility supply water temperature, with W1 class with the most restrictive margin (35.6–62.6 °F) and W5 class with the least restrictions (113 °F).

Another one of very important considerations when water-cooled equipment are selected for the data center is that due to the temperature difference between the cooling liquid inside the pipes and environment temperature restricted insulating provisions should be considered to prevent possibility of any condensation over the pipes cold surfaces.

Despite of which type of cooling system is designed for the data center, the air distribution to the computer room is usually through the bottom discharge air handling units which discharge the air into an under-floor plenum space. Air then will be entered to the space through heavy duty floor mounted supply devices. Properly locating the air temperature sensors (usually at the level where they enter to the racks) can be very effective in efficient operation of the system. The minimum recommended height of the raised floor is usually about 24″ in order to help creating enough space for both air and cables and cable trays and therefore allowing uniform air pressure for the whole computer room. The configuration of the data racks' supply and return paths are designed such that the cold air enters from the under-floor plenum to the cold Aisle(s) in front of rack(s), moves through the racks to the hot Aisle(s) in the back of the racks and leaves the room through the above the ceiling return air plenum path back to the air handling unit. A simple illustration of this air moving path has been shown in Fig. 3.1 below. If the engineer expect that this air becomes hot enough to be returned to the air handling unit, this air may not be returned to the air handling unit and instead be used somewhere in the facility for preheating purpose. Such condition then forces utilization of 100 % outdoor air units capable of operating without return air path to the air handling unit. Sometimes blocking the space between the top of the computer and the ceiling that makes complete separation between hot and cold Aisles can improve the efficiency of the system considerably (see Fig. 3.2).

Most popular and efficient units utilized in small computer rooms are air- or water-cooled DX (Direct Expansion) units with airside economizer cycle. Water-cooled DX units utilize cooling tower, condenser water pumps and plate and frame heat exchanger as well. On the other hand, for large computer room units the most efficient systems are usually high efficiency water-cooled chillers equipped with variable frequency drive (VFD). When computer room is located in a location with

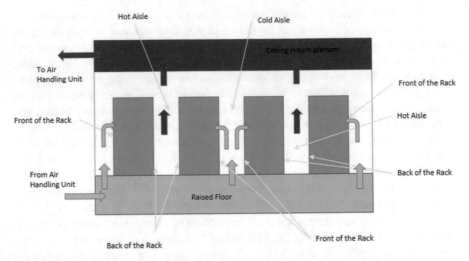

Fig. 3.1 Typical data center hot–cold aisle strategy

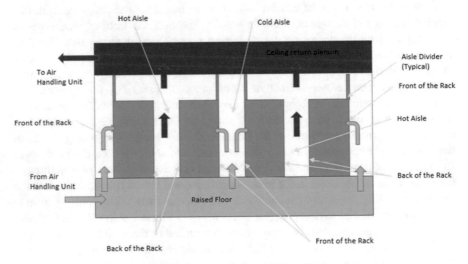

Fig. 3.2 Typical data center hot–cold aisle strategy with aisle dividers

dry climate implementing water side economizer can create higher efficiencies specifically where a large difference between maximum and minimum load conditions is expected.

One of the good general practices in designing energy efficient data centers is to use multiple smaller cooling equipment (e.g., chillers, or DX units) instead of a single large unit. Such provision will allow to reduce the power input to the data center when it is operating in part load conditions, and also would contribute to the future expansions of the data center by adding more small units to the site as required to satisfy the gradual expansion of the data center and of course without disturbing the existing system operation [4].

ASHRAE [3] has offered a series of recommendations for designing energy efficient data centers. A range of tips from optimizing supply air temperature, optimizing pumping control, optimizing IT equipment cabinet airflow, optimizing humidity control, improving computer room unit controls, improving return air management, and improving under floor pressure management are suggested in this reference.

ASHRAE [4] best practice guide covers a range of recommendations for efficiency improvement in data centers such as defining specific ways for improvement in environmental criteria (e.g., implementing control strategies to prevent different cooling units fighting), and ways for minimizing mechanical equipment and systems energy consumption (e.g., using load matching techniques) and of course commissioning.

ASHRAE Application Handbook [5] defines a performance indicator that can be used for comparing overall energy efficiency of different data centers. Power usage effectiveness (PUE) is defined as a ratio of total data center energy consumption or power to the energy consumption or power of information technology (IT) equipment, where the IT energy consumption should be measured for the IT equipment only.

Eubank et al. [6] as the result of their charrette that was performed in 2003 and brought together about 90 high-level data center design and management experts in order to recommend best solutions for data center design reported a few important outcomes of their exercise which some of the most important of these recommendations are as follows: (1) the value of saving 1 W compounds quickly throughout the total data center system, (2) looking only at energy consumption does not disclose overall operational costs, and (3) efficient design massively reduces the quantity of material resources needed to provide computing services.

For a thorough and in-depth discussion for learning different concepts associated with the data center and also designing the HVAC systems suitable for them refer to ASHRAE Application Handbook [5], and other references [2–4] and [6] noted earlier in this chapter.

References

1. U.S. Department of Energy. (2011). Best practices guide for energy-efficient data center design. Federal Energy Management Program. Prepared by the National Renewable Energy Laboratory (NREL), a national laboratory of the U.S. Department of Energy, Office of Energy Efficiency and Renewable Energy; NREL is operated by the Alliance for Sustainable Energy, LLC.
2. ASHRAE Datacom. (2012). *Thermal guidelines for data processing environments* (ASHRAE Datacom series, Book 1 3rd ed.). Atlanta, GA: American Society of Heating, Refrigerating, and Air Conditioning Engineers.
3. ASHRAE Datacom. (2010). *Green tips for Datacom Centers* (ASHRAE Datacom series, Vol. 10). Atlanta, GA: American Society of Heating, Refrigerating, and Air Conditioning Engineers.

4. ASHRAE Datacom. (2009). *Best practices for Datacom facility energy efficiency* (Datacom series 2nd ed., Vol. 6). Atlanta, GA: American Society of Heating, Refrigerating, and Air Conditioning Engineers.
5. American Society of Heating, Refrigerating and Air Conditioning Engineers. (2015). *ASHRAE application handbook*. Atlanta, GA: American Society of Heating, Refrigerating and Air Conditioning Engineers.
6. Eubank, H., Swisher, J., Burns, C., Seal, J., Emerson, B. Design recommendations for high-performance Data Centers, Report of the Integrated Design Charrette, Conducted 2–5 February 2003, Rocky Mountain Institute (RMI), with funding from Pacific Gas & Electric Energy Design Resources program. www.rmi.org/sitepages/pid626.php

Chapter 4
Health-Care Facilities

Abstract US Department of Energy report shows the inpatient medical facilities (2.45 billion square foot) in 2015 use 0.51 quadrillion BTUs of energy with an intensity of 208.2 thousands BTUs per square foot. With such a high level of usage and intensity it can easily be claimed that the health-care industry is one of the most energy-consuming applications in building industry.

Keywords Health-care facility • Pressurization • Economizer • HEPA filter • Surgery • Humidification • Dehumidification • Energy recovery system • Energy intensity

US Department of Energy report http://buildingsdatabook.eren.doe.gov/Table View.aspx?table=3.8.2 shows the inpatient medical facilities (2.45 billion square foot) in 2015 use 0.51 quadrillion BTUs of energy with an intensity of 208.2 thousands BTUs per square foot. With such high level of usage and intensity it can easily be claimed that the health-care industry is one of the most energy-consuming applications in building industry. Due to the importance of this application which has direct effect on the wellness of people, multiple codes and standards such as "The American Institute of Architect's Guidelines for Design and Construction of Hospital and Health Care Facilities (AIA Guidelines)," "Standards and handbooks of the American Society of Heating, Refrigerating and Air-Conditioning Engineers (ASHRAE)," and "Centers for Disease Control and Prevention (CDC) guidelines and recommended practices." have developed restricted codes and guidelines for designing health-care facilities. ASHRAE [1] defines the duty of the HVAC system in a health-care facility in addition to facilitating the general comfort (temperature, humidity, air movement, noise, etc.) level for the occupants, as one of the premier means for the environment control against infection, control for specific medical functions, hazard control and life safety measures. The HVAC system can contribute to infection and hazard control in the building by diluting the air (exhausting the contaminated air and replacing it with clean fresh air), directional airflow from clean spaces to dirty spaces before exhausting, using high-efficiency filtration (e.g., 90–95 % filters in patient rooms and 99.97 % (HEPA) filters in protective environment and specialty operating room—see Fig. 4.1 below), and using ultraviolet germicidal irradiation in air handling units to eliminate airborne microorganisms and decrease the patient risk

© Springer International Publishing Switzerland 2016
J. Khazaii, *Advanced Decision Making for HVAC Engineers*,
DOI 10.1007/978-3-319-33328-1_4

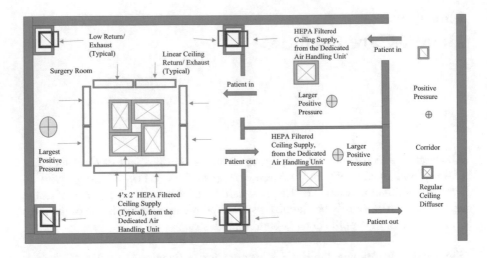

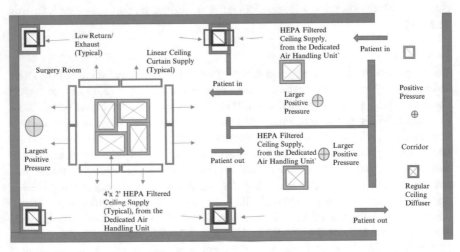

Fig. 4.1 (**a**) and (**b**) Typical relative pressure control in surgery room in health-care facilities [4]

of infection. Each of these duties can contribute to increasing the energy consumption of the system dramatically. As a result of this great energy consumption potential in the health-care facilities, implementing as much as energy-saving measures as possible such as utilizing heat recovery systems, installing economizers, applying demand shifting strategy, and employing hybrid cooling methods should always be considered and executed where it is possible.

ASHRAE Standard 170 [2] defines the ventilation rate requirement for health-care facilities. The fact that each space has to be provided with a specific minimum quantity of air, a specific minimum air change and yet it should be exactly pressurized either positively or negatively in relation to the adjacent spaces makes the outdoor air ventilation a very large part of the energy consumption of health-care facilities. Where in a space is required to maintain the positive or negative pressure, the return air should have either motor operated damper or a variable air volume unit to track the supply air variable air volume unit to keep the constant air pressure difference in the space. The minimum outdoor air required for each space is not only defined based on the number of people, size of the space, minimum air change, minimum outdoor air required but also can be affected by the presence of appliances such as hoods or safety cabinets in the space. So if in a space there is a hood or even two or three hoods, the designer has to allow for enough outdoor air to this space so even when all the hoods are operating simultaneously there is still enough outside air available to keep the correct relative pressure for the space regarding to the adjacent spaces. It can be seen that as we move room to room and attempt to design the proper outdoor air provisions for them, it could add up to a very large outdoor quantity requirement very soon.

There are plenty places all over a health-care building with different types of hazardous contaminants that are required to be exhausted, but it is not usually practical or even allowed that these different types of exhausts to be connected to a central exhaust system. Exhaust from rooms such as ER waiting, triage, radiology waiting, and bronchoscopy shall be ducted separately and exhausted separately with all the required and precautionary recommendations in relevant codes. Therefore, the chance for energy recovery through the exhaust air is usually just limited to general type exhausts from restrooms, locker rooms, and similar spaces. Another fact is that in health-care applications usually the air handling unit coils should be designed large enough so even if due to any reason the energy recovery unit fails, the main system can operate fully without counting on the preheating or precooling that was supposed to be supplemented through energy recovery units.

In a report resulted from research in different climate zones in the USA performed by ASHRAE [3] a few different systems for providing air conditioning for large hospitals were evaluated with a target of reducing significant energy consumption where it is possible. At the end of this research the recommended cooling systems were distributed water source heat pumps with designated outside air unit, fan coil units with designated outside air units and mixing variable air volume systems with separated outside air treatment and heat recovery for noncritical care spaces. For critical care spaces such as surgery room, trauma room, and intensive care room. ASHRAE [3] also recommended central air handling unit (water-cooled chiller, variable frequency drive water circulation pumps, cooling towers with variable frequency drive for fan, hot water boiler and economizer). It also was noted that some regions with more restricted laws may demand dedicated constant air volume systems in critical care rooms.

Probably one of the unique design challenges associated with the health-care facilities is HVAC design for the operation room. There have been different approaches for configuration of the airflow in such spaces. In different designs, almost exclusively the majority of the air is delivered through HEPA filtered laminar flow panels right above the operation table, and the exhaust/return air is to exit the room through low sidewall return/exhaust devices. The difference between the approaches is where some researchers prefer returning part of the return/exhaust from the room from linear return devices surrounding the central clean space, while the other group prefer providing another supply curtain to separate the central laminar flow space from the rest of the operation room (see Fig. 4.1a, b, that are drawn based on a figure represented in ASHRAE Journal [4]). Either approach that is chosen by the designer should strictly follow the applicable codes and standards in the region of design.

In a report in 2010 [5] researchers represented that for two general prototype hospital designs with three different approaches in HVAC system, a reduction of energy intensity (EI) for a typical operational hospital from 270 kBtu/ft^2.year to almost 100 kBtu/ft^2.year is achievable. The main energy reduction was produced by improving the space heating and hot water production systems by reducing it by 92 % which changed the energy source use balance from common 40–60 % (electricity–natural gas) to 82–18 % (electricity–natural gas) ratio as a consequence. The proposed HVAC system options other than the baseline code compliance option were conventional central energy recovery system, and large ground source heat pump system. It was also shown that a combination of different strategies such as decoupling of cooling, heating and ventilation systems, use of decentralized systems such as radiant panels chilled beams and fan coil units where possible, and day lighting helped this goal to be achieved.

Due to the importance and high dependency of the occupants health of the HVAC system many parts of the HVAC system shall be designed with not only emergency power but also with up to 100 % redundancy to eliminate the chances of losing their functionality due to either loss of power or need for maintenance where they are required to operate.

Many codes and standards identify and regulate typical rooms that need to be strictly exhausted due to the type of contamination stored in them such as soiled linen rooms and waste storage rooms. These codes and standards identify and regulate some equipment that are required to be exhausted also due to the high level of existing hazardous material processed in them such as exhaust from radioisotope chemical fume hoods. These hoods should also be equipped with HEPA filters at discharge to prevent the outside environment from being exposed to these contamination.

ASHRAE [1] uses the model of a 350,000 ft^2 full service metropolitan health-care facility with all the typical health-care facility functions, which from the percentage energy consumption end users are very close to DOE-2 selected models for 750,000 ft^2 suburban health care, 260,000 ft^2 regional health-care facility, and 175,000 eye institute/general hospital models. The outcomes of modeling with all these prototypes show the importance of the HVAC system in energy consumption of the whole facility due to: "First, because of the significant minimum total air change requirements and minimum outdoor air change code requirements for health care facilities, between 31 and 41 % of the annual energy cost of the typical facility is used for ventilation fan energy and for overcooling and then reheating space ventilation air in excess of that required to heat or cool the space. Additionally, another 13–20 % of the annual energy cost of the facility is used for auxiliary or "parasitic" energy usage components, composed of pumping energy and cooling tower energy. Between 5 and 10 % of the annual energy costs are required for outdoor air heating, cooling, humidification, and dehumidification, depending upon location" [1].

Similar to other applications, some effective approaches to reduce energy consumption of the HVAC systems in health-care facilities are, use of well-designed variable volume air handling units where codes and specific application permit, and taking advantage of allowed lower air circulation in unoccupied noncritical spaces, and also proper zoning. Using a variable volume system generally helps to modulate down the airflow volume in space where there is no need for full air conditioning due to the part load conditions. This is not a usual option for health-care facilities critical spaces specifically when we need to maintain the relative pressure of the space in a certain level at all time. But it can be helpful when the noncritical space is not in full load conditions or is in unoccupied mode [6]. Because of the nature of the health-care facility there is a relatively larger demand for air quantity and movement in this type of application compared to others. Different literatures claim that in general almost half of the pressure loss in health-care facility air conditioning units are due to the air handling unit ductworks and the other half is due to the air handling own internal losses. A very useful method of energy saving other than having a proper duct layout, therefore, is to decrease the air speed inside the air handling unit and cross the coil surfaces. The common value for air flow velocity over the coil surface for general purpose air handling units is around 450–500 ft/min. This value in some health-care facility can be reduced to near 250 ft/min. This requires a larger cross section of air handling unit in mechanical room, but reduces the fan power dramatically which as a result would be a considerable energy-saving provision.

The other useful approach in order to minimize the energy consumption in typical health-care facilities is proper zoning [6]. In every health-care facility some administrative and general areas exist that not only do not require special outdoor air quantity above the general office requirement, but also its relative pressure do not need to be controlled closely either. This provides an opportunity to assign these spaces that have similar function to their own zones and avoid mixing their treatment with specific critical mission spaces and therefore air handling units and cooling associated with them.

Other options for energy saving in health-care facilities are included using different technologies of heat recovery such as heat pipe and heat wheel, designing energy efficient chilled water and condenser water systems such as water side economizers, variable speed chillers, pumps and cooling tower fans, and of course energy efficient heating systems such as use of high efficient condensing boilers.

It can be seen and as noted earlier heating for comfort as well as producing hot water for domestic use and therefore heating plants efficiency in health-care facilities can have great importance. Health-care facilities operates every day on a 24-h basis. Patients sleep in health-care facility, and therefore, they need heat available for comfortable condition, most of the time steam is required for sterilization purposes, hot water is needed for clinical functions as well. Available and sufficient heating for critical care spaces such as operating rooms, delivery rooms, recovery rooms, and intensive care rooms are vital part of the health-care facility operations. This calls for sufficient backup boilers, boiler feeds, pumps, etc. and sufficient emergency power to make sure in occasions of system failure or regular maintenance none of these duties are endangered [6].

Most of the health-care facilities are equipped with some sort of laboratories. Laboratories are another source of consuming large quantity of energy and due to its importance I dedicate the following chapter in this book to discussing laboratories as a whole independent entity, which could also be considered a part of the health-care facility.

For a thorough and in-depth discussion for learning different concepts associated with the health-care facilities and also designing guidelines for HVAC systems suitable for them refer to ASHRAE Application Handbook [6] and also other referenced material in this section.

References

1. ASHRAE. (2003). *HVAC design manual for hospitals and clinics*. Atlanta, GA: American Society of Heating, Refrigerating and Air-Conditioning Engineers, Inc.
2. American Society of Heating, Refrigerating and Air-Conditioning Engineers, Inc. (2013). *ANSI/ASHRAE/ASHE Standard 170 - 2013*. Atlanta, GA: American Society of Heating, Refrigerating and Air-Conditioning Engineers, Inc.
3. ASHRAE, AIA, IESNA, USGBC, USDOE. (2012). Advanced energy design for large hospitals. American Institute of Architects, Illuminating Engineering Society of North America, US Green Building Council, US Department of Energy.
4. Drake, B. (2006, June) Infection control in hospitals. *ASHRAE Journal, 48*, 12–17.
5. Burpee H., & Loveland, J. (2010) Targeting 100! Envisioning the high performance hospital: Implications for a new, low energy, high performance prototype. University of Washington's Integrated Design Lab with support of Northwest Energy Efficiency Alliance's (NEEA) BetterBricks Initiative in collaboration with NBBJ Architects, Solarc Architecture and Engineering, TBD Consultants, Cameron MacAllister, Mahlum Architects, Mortenson Construction.
6. American Society of Heating, Refrigerating and Air Conditioning Engineers. (2015). *ASHRAE application handbook*. Atlanta, GA: American Society of Heating, Refrigerating and Air Conditioning Engineers.

Chapter 5
Laboratories

Abstract Another application in the group of large HVAC energy consumers is laboratory. Laboratories can be designed as a stand-alone facility or as a part of a larger health-care facility. In this chapter I discuss the energy related issues and helpful provisions to improve the energy efficiency of the laboratories from the view point of designing HVAC system for a stand-alone laboratory. Most of these provisions can be easily implemented in the study of health-care systems and therefore become a complementary to the provisions described in the previous chapter (health-care facilities) where laboratories are built as a part of the general health-care facility.

Keywords Laboratories • Efficiency • Chemical • Research • Sash • Hood • Cabinet • Dilution • Effectiveness • Heat pipe

Another application in the group of large HVAC energy consumers is laboratory. Laboratories can be designed as a stand-alone facility or as a part of a larger health-care facility. In this chapter I discuss the energy related issues and helpful provisions to improve the energy efficiency of the laboratories from the view point of designing HVAC system for a stand-alone laboratory. Most of these provisions can be easily implemented in the study of health-care systems and therefore become a complementary to the provisions described in the previous chapter (health-care facilities) where laboratories are built as a part of the general health-care facility. It should be noted that laboratories can be built as a part of other facilities such as educational buildings as well.

Similar to the general flow of the book and what has been discussed in previous chapters and also since laboratories are big energy consuming application, it can be understood easily that the proper design of its HVAC system can play a major role in efficiency improvement of the whole application. In addition to the general complications associated with any energy consuming applications the laboratories are specifically challenging to design due to their variety of types. A laboratory can be one of many types such as general chemistry, radiology, teaching, research, hospital, biological, and animal. Each type will be required to satisfy a specific function and therefore requires different HVAC design and treatment. For example a general chemistry laboratory usually is targeted for performing general experiments with a variety of chemicals (not for extremely dangerous chemicals) which

© Springer International Publishing Switzerland 2016   45
J. Khazaii, *Advanced Decision Making for HVAC Engineers*,
DOI 10.1007/978-3-319-33328-1_5

the only laboratory specific equipment in this room might be one or a few chemical fume hood(s), while a research laboratory usually is focused on performing experiments with a wide range and large quantity of material (including dangerous chemicals) and therefore requires much more accurate environment control and also high end instruments in addition to general fume hoods. Further in an animal laboratory safe and of course separated environment for animal and laboratory personnel is the priority of design. Other types of laboratories such as isolation (cleanroom) and electronics laboratories also exist [1]. Since cleanrooms have such extreme requirements in quantity and also velocity of the introduced air to the room, I have designated a separate chapter in this book for discussion about them.

Designing a laboratory is a very complicated task and requires many years of experience along with complete familiarity with multiple codes and standard to be strictly followed. ASHRAE Applications Handbook [1] has presented a long list of resource materials that are applicable to different types of laboratories. To name a few are laboratory ventilation [2] by AIHA, health-care facilities [3], fire protection for laboratories using chemicals [4] and fire protection guide for hazardous materials [5] by NFPA, Class II biosafety cabinetry [6] by NSF, and biosafety in laboratories [7] by NRC.

In fact what in general makes the laboratories a great energy-consuming application is the need for performing some experiment under controlled environment. Examples of such experiments are performing chemical reactions under fume hoods or performing research on specific toxic material under biosafety cabinets. In order to keep the person who is performing the experiment and also other people in the room and even building from possible danger of being exposed to undesired and dangerous experiment material or products usually a large quantity of conditioned air should be pushed through and being exhausted by the hoods and cabinets and its associated exhaust fans [1].

The most popular equipment type in general laboratories are chemical fume hoods. They are used to protect the personnel by providing an inward airflow through the hood sash (movable front opening glass). In most general configuration (conventional fume hood) an exhaust fan exhausts the air that enters to the hood cabinet space. Hood exhaust fan airflow is set based on the desirable air velocity at the cabinet opening to keep the gas products of the experiment from pushing back and entering the laboratory space. The sash position can be changed from fully open to fully closed and therefore the air velocity at the entrance to the cabinet will change accordingly to match the constant exhaust fan air flow. In order to manage the air velocity changes at the entrance to the fume hood that may result in a higher than desirable air velocity and therefore poor protection of the personnel working in front of the hood, different fume hood configuration may be utilized. Fume hoods such as bypass type, baffle type, variable air volume type (see Fig. 5.1) and auxiliary type each are designed to somehow contain the airflow velocity at the entrance to the hood at an optimum level as the sash position and therefore the entering air surface to the hood cabinet changes. In some occasions multiple fume hoods will be connected to a general exhaust manifold and will be discharged through an (a set of) exhaust fan(s). In such conditions and in order to keep the face

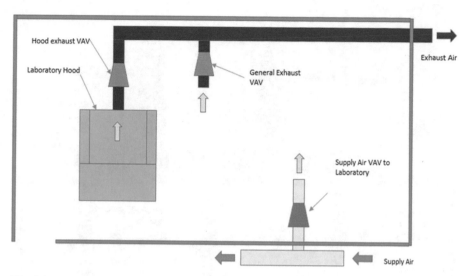

Fig. 5.1 Typical variable air volume system strategy for laboratories

velocity of all the chemical fume hoods at the desired level, the exhaust fan(s) will be designed at a dilution fan configuration (see Fig. 5.2). A dilution fan is a constant flow fan which at all time exhaust the same airflow, but as the position of the sash (es) change(s) and therefore airflow into the hood(s) change(s), dilution/bypass dampers on roof modulate to allow the difference required for air to get to the exhaust fan. So the fan will utilize a portion of outside air as the complement and additional air for constant operation while the air velocity at the entrances to the fume hoods remains at the desired level.

Next to the general chemical fume hoods, biological safety cabinets are the other most common equipment used in laboratories. As it is obvious from its name this type of cabinets is commonly being used for experiment with the biological and toxic agents. They are divided to Type 1, Type 2 (A, B1, B2, & B3) and Type 3 biological safety cabinets [1]. The division is based on the level of toxicity and danger of the substances that are subjected to be tested or manipulated there, with Type 1 being suitable for the least and Type 3 being suitable for the most toxic materials. Type 2 is subdivided to four different categories by itself and the differences here are based on the amount of recirculated air from the exhaust path back to the cabinet enclosure which by itself is depended on the level of toxicity of the substances. All the exhaust and recirculated air to the cabinet enclosure need to be provided with HEPA filters, while only Type 3 biological safety cabinets have to be equipped with intake HEPA filter as well. Velocity of the flow at intake should be controlled and that will make these cabinets a source of energy consumption.

Another type of equipment that are typically used in laboratories are storage cabinets. They are utilized to store and keep highly flammable or solvent materials and are designed to prevent exposure of these material to condition that may cause fire or explosion by means of venting (if required) and separation from the general laboratory environment.

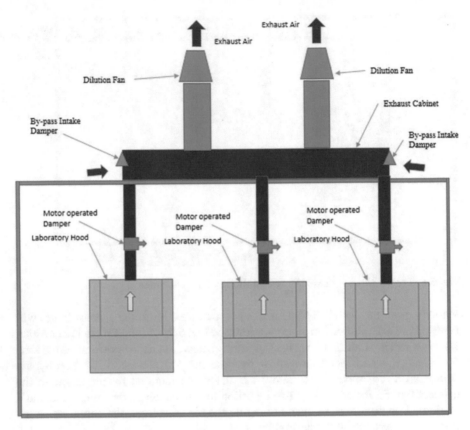

Fig. 5.2 Dilution fan assembly for multiple laboratory hoods

The existence of multiple sources of large volume of exhaust and also the need for directing the airflow to a specific direction in order to keep the proper pressure in different laboratory spaces demand a large quantity of outdoor air (most of the time almost 100 % outside air) to be treated and delivered to the building at each instance. That is a great source of energy consumption which with good engineering design can also become the source for implementing energy saving measures as well. Other than selecting the general equipment with highest possible efficiencies, probably the only means for energy saving in such large energy consumers is the utilization of some sort of energy recovery unit. Energy recovery units can be one of the air-to-air, water-to-air, or hot gas-to-air recovery unit types. Some of the most popular energy recovery approaches in laboratory design are using enthalpy wheel, run-around loops and heat pipes that are also popular remedies in energy saving in other applications as well [1]. The effectiveness of an enthalpy wheel can be as high as 75 % where sensible effectiveness of heat pipe is usually between 45 and 65 %. The sensible effectiveness of run-around loop is usually between 55 and 65 % as well. When the cross contamination can be a possible problem, heat pipe and run-around coils are better alternatives for heat recovery from the exhaust of the laboratory systems, specifically since in this case the NFPA 45 only allows heat recovery to be done on general exhaust. The advantage of a run-around coil is that contrary to energy wheel and heat pipe it does not require the supply and exhaust air ducts be near each other.

ASHRAE Standard 90.1 [8] requires that buildings with more than 5000 cubic feet per minute laboratory exhaust systems to have at least one of the following features. "(1) Both supply and exhaust systems having variable volume controls and be capable to of reducing exhaust and makeup or install a heat recovery unit that fulfill the following condition (sum of percentage that the exhaust and makeup airflow rates can be reduced from design condition plus multiplication of the division of exhaust airflow rate through the heat recovery device at design condition by the makeup air flow rate of the system at design condition multiplied by percentage sensible recovery effectiveness be at least 50 %), (2) Variable air volume supply and exhaust systems that can keep the required minimum circulation standards or pressurization, and (3) direct makeup air supply of at least 75 % of the exhaust capacity given the specified condition in the standard [8]."

ASHRAE Application Handbook [1] also proposes searching and determining the possible opportunities during unoccupied periods, and managing an unoccupied setback approach by differentiating between the flow rates between occupied and unoccupied periods and specifying trigger points that make switching between setback and normal operation possible.

I close the section with an important quote from ASHRAE Applications Handbook [1]: "Laboratories present unique challenges and opportunities for energy efficiency and sustainable design. Laboratory systems are complex, use significant energy, have health and life safety implications, need flexibility and adaptability, and handle potentially hazardous effluent with associated environmental impacts. Therefore, before implementing energy efficiency and sustainability design protocols, the engineer must be aware of the effects of these measures on the laboratory processes, which affect the safety of the staff, environment, and scientific procedures."

For a thorough and in-depth discussion for learning different concepts associated with the laboratories and also designing the HVAC systems suitable for them refer to McIntosh [9] and ASHRAE Application Handbook [1].

References

1. American Society of Heating, Refrigerating and Air Conditioning Engineers. (2015). *ASHRAE application handbook*. Atlanta, GA: American Society of Heating, Refrigerating and Air Conditioning Engineers.
2. AIHA. (1992). *Laboratory ventilation, ANSI/AIHA Standard Z9.5*. Fairfax, VA: American Industrial Hygiene Association.
3. NFPA. (1996). *Healthcare facilities. ANSI/NFPA Standard 99*. Quincy, MA: National Fire Protection Association.

4. NFPA. (1986). *Fire protection for laboratories using chemicals. ANSI/NFPA Standard 45*. Quincy, MA: National Fire Protection Association.
5. NFPA 491M. *Fire protection guide for hazardous materials*. National Fire Protection Association, Quincy, MA.
6. NSF. (2008). Class II biosafety cabinetry; Design, Construction, Performance and field Certification. NSF International Standard American National Standard 49.
7. NRC. (1989). *Biosafety in laboratories; prudent practice for handling and disposal of infectious materials*. Washington, DC: National Research Council, National Academy Press.
8. American Society of Heating, Refrigerating and Air Conditioning Engineers. (2013). *ASHRAE Standard 90.1 Standard. Energy standard for buildings except low-rise residential buildings*. Atlanta, GA: American Society of Heating, Refrigerating and Air Conditioning Engineers.
9. McIntosh, I. B. D., Dorgan, C. B., & Dorgan, C. E. (2001). *ASHRAE Laboratory design guide*. Atlanta, GA: American Society of Heating, Refrigerating and Air Conditioning Engineers, ASHRAE.

Chapter 6
Cleanrooms

Abstract A cleanroom is a manufacturing enclosed space in which high quality, high technology products such as semiconductors, pharmaceuticals, aerospace electronics, and photovoltaic cells are produced. ASHRAE Application Handbook (ASHRAEApplication Handbook, ASHRAE handbook, HVAC applications. American Society of Heating, Refrigerating and Air conditioning Engineers, Atlanta, GA, 2015) divides the cleanroom applications into pharmaceutical (biotechnology), microelectronics (semiconductors), aerospace, and other miscellaneous applications such as food processing and manufacturing of artificial limbs. The most critical focus in such spaces are not the usual requirement of the typical highly controlled temperature, humidity, air pressure, etc. but it is the need for tight controlling and effectively removing particles from the room.

Keywords Cleanroom • Pharmaceutical • Biotechnology • Microelectronics • Semiconductor • Flakes • Unidirectional • Redundant • Tools

A cleanroom is a manufacturing enclosed space in which high quality, high technology products such as semiconductors, pharmaceuticals, aerospace electronics, and photovoltaic cells are produced. ASHRAEApplication Handbook [1] divides the cleanroom applications to pharmaceutical (biotechnology), microelectronics (semi-conductors), aerospace, and other miscellaneous applications such as food processing and manufacturing of artificial limbs. The most critical focus in such spaces are not the usual requirement of the typical highly controlled temperature, humidity, air pressure, etc. but it is the need for tight controlling and effectively removing particles from the room. This requirement is due to the vital effect of pollution particles on performance of high quality products such as semiconductors and photovoltaic cells if in some way pollution reaches and infects the product [1]. Therefore, the first thing to consider when designing a cleanroom is to understand what the sources of contamination are and find proper ways to prevent them to get into the space. Also if these pollutions somehow enter the space or even are generated inside the space finding solutions to remove them from the cleanroom is a basic requirement. We can easily imagine that the sources of contamination for a cleanroom can be divided to two external and internal sources. The main source of external contamination is outdoor air. Outdoor air can find its way to the inside of the cleanroom via outside air introduced by air handling units, infiltration through

© Springer International Publishing Switzerland 2016   51
J. Khazaii, *Advanced Decision Making for HVAC Engineers*,
DOI 10.1007/978-3-319-33328-1_6

cracks, and opening the doors. One of the solutions to protect against these possible sources of contamination is usually providing high efficiency filters at air handling unit intake and sometimes even at each single air discharge diffuser. Another possible solution is pressurizing the cleanroom to prevent negative air pressure inside the cleanroom and therefore decreasing the tendency of the pollution in the outdoor air to push inside the space. Also a very tight envelope structure can prevent infiltration through building cracks. The sources of internal contaminations are usually people who are working in the cleanroom and their residuals such as skin flakes, sneezing, and cosmetics. In addition construction material such as finishes, paint, and also products of combustion, and manufacturing processes are other main sources of internal contamination in a cleanroom. The general methods for controlling the internal sources of contamination is to select cleanroom proper construction material for the interior side of the building, proper clothing for the people entering and working in the cleanrooms and air locks between the cleanroom and the exterior environment [1].

Different classes of cleanroom have been defined by ASHRAE Application Handbook [1] based on the size of the particles allowed in the space. Tables that are used for this purpose show the acceptable number of particle in each cubic foot volume of the cleanroom for each of the specified particle sizes. Similar tables show the minimum and maximum recommended air change and air velocity for each class of cleanrooms. For example based on ASHRAE Application Handbook [1] for a class-6 cleanroom the maximum allowed airborne particle concentration for particles of size 0.5 μm is 35,200 particles per 1 cubic meter of cleanroom space, and for the same class cleanroom the maximum allowed airborne particle concentration for particles of size 0.1 μm is 1,000,000 particles per 1 cubic meter. In order to keep this level of cleanness for a cleanroom for example with a height of 40 ft, the vertical velocity of the air should be kept between 25 and 35 ft/min. That requires equivalent of 38–53 air changes per hour in this cleanroom. Similarly the recommended air velocity and air changes for a class-2 cleanroom with height of 40 ft, will be between 85 and 100 vertical feet per minute and between 125 and 150 air changes per hour. Considering the large foot-print of the manufacturing sites (cleanrooms) for manufacturing semi-conductors, photovoltaic cells and other critical products we can imagine the huge amount of supply air that would be required to be delivered to the cleanrooms. Additionally, and depending on the required exhaust airflow capacity out of cleanroom which is due to different manufacturing tools producing unclean exhaust products a very large portion of this supply air to the cleanroom has to be provided through utilization of the treated and conditioned outdoor air to keep the correct level of pressurization inside the cleanrooms. This is why the magnitude of energy consumption for cleanrooms tends to become outrageously huge. Not to mention that most of the exhaust flow from the site are required to move through some sort of cleaning process or use of specialized scrubbers before being discharged to the outside environment. As the result of the effect of these additional processes and scrubbers exhaust fans are usually much larger and consume much more energy compare to general exhaust procedure in the other applications.

Most of the modern manufacturing clean plants use vertical unidirectional air supply to the cleanroom. Through either one of the filtered pressurized plenum method, filtered ducted or filtered individual fan-powered method supply air is introduced to the room. Considering the quantity of required supply air as it was discussed earlier it is not unusual to have situations that almost all the ceiling of the cleanroom to be used as supply outlets. All the outlets are covered with HEPA filters to prevent introduction of the particles to the plant. Return air will be sent back to the air handling unit through either floor and under-floor return plenum or through side wall return outlets and paths. This configuration helps a complete and uniform washout of the whole cleanroom from top to bottom and prevents the particles from contaminating the products of the cleanroom manufacturing process. In some applications horizontal air delivery may be considered. In this approach air is forced to move from one side of the cleanroom to the other side while the pattern of locating the manufacturing tools in the space and therefore direction of air is designed to push air over the cleanest operations first, and leave the room from the end which contains the least clean manufacturing process tools [1].

The air delivery usually will be done in two steps. In first step multiple dedicated outside air units will be designed to pretreat the outside air and makes it ready to be delivered to the main circulating air handling units. The next set of air handling units receive this treated air and circulates it inside the cleanroom with minimum of cooling, reheating, humidification and dehumidification when it is required (Fig. 6.1).

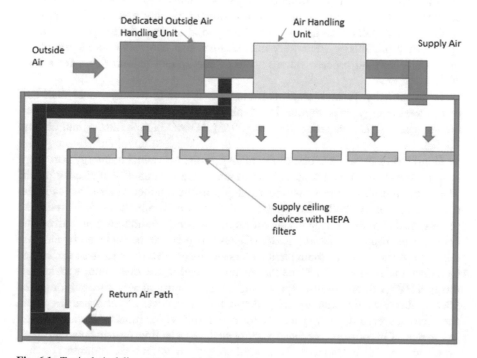

Fig. 6.1 Typical air delivery strategy to cleanroom

Another source of energy consumption in cleanrooms is usually the need for large quantities of process cooling water. A large number of tools in cleanroom are usually water-cooled equipment which means they need cold water to transfer the generated heat inside the production tool out of the system. Therefore, the cooling plants (chillers) should be not only capable of cooling a significant quantity of outside air as it was discussed earlier but also should be capable of producing enough cooling to remove the heat that is generated from water-cooled tools that are used inside the cleanroom. These two major cooling requirements call for relatively very large chiller sizes compared to the other applications.

Many manufacturers are not willing to undergo a down time in their production due to the possible huge financial hit that they may be faced due to lack of production. Therefore, due to this manufacturer's lack of tolerance and also extensive potential damages to both site and environment, if production operation continues without the proper preventive functions in case of system failure even for a short period of time, almost all of the cooling sources (specifically process water cooling system) and also all the exhaust systems should be designed with full redundant capacity. For example an exhaust system designed to remove acid exhaust from one operating tool should be consisted of two full size exhaust fans (one operating and one redundant) connected parallel to the exhaust duct that is carrying the acid exhaust. This configuration helps the manufacturing process continues even when one of the exhaust fans is not being capable of operation, due to either fan failure or routine maintenance. Another example is the dire need for a full redundant process water cooling system consisted of additional source of cooling (redundant chillers), and also additional pumping system. If the process cooling system fails even if it is for a short period of time, the temperature in the system may potentially reach very high degrees that can melt down some of the pipes carry cooling process water and force the system to shut down for a long period of time.

A research by Tschudi et al. [2] represented that in average more than half of the total energy consumption in a typical cleanroom comes from its environmental systems, i.e., cleanrooms' fans (30 %), HVAC pumps (20 %), and chillers (14 %).

Considering these facts one of the main options for reducing energy consumption in cleanrooms must be through smart managing the quantity of circulating air. The circulated air quantity has direct relation to the number of air changes in the space, and consequently the physical size of the room. Therefore, reducing the physical size of the cleanroom within the possible limitations could affect the energy consumption in the room. Sometimes this goal can be achieved by creating dummy spaces inside the main plant and locate the process tools in these enclosures to reduce the required air. Since the air introduced to the cleanroom should pass through HEPA filters, smaller spaces and lower quantity of air passing through the HEPA filters would help to bring down the associated air pressure drop and therefore decrease the power and motor size required for running the air through the system. Of course other usual factors such as selecting proper fans with high efficiency would contribute to the overall energy saving in the cleanroom.

Even though it looks like the other opportunity for decreasing the energy consumption in cleanrooms is decreasing the quantity of outdoor air and therefore decreasing the capacity of cooling that is required, but the quantity of the outdoor air has direct relation to the amount of process exhaust that leaves the space, and it is not completely open for designer to work around that. To calculate the exact quantity of outside air the designer should add proper quantity of air required for the pressurization of the space to the total exhaust capacity of the air from the space. The exhaust capacity should be counted by summing up the maximum possible process exhaust from different tools or other general sources of exhaust in the space.

As it was noted earlier another important source of energy consumption and therefore opportunity for saving energy while designing a cleanroom is the efficient central plant consisted of chillers and pumps. In order to have an efficient central plant common factors such as use of free cooling (air side economizer) where it is possible, selecting full load and part load efficient chillers and pumps, minimizing the pressure and flow resistance by utilization of smart piping layout, and optimizing chilled water temperature should be considered. Another consideration for the cleanroom design can be separating the dedicated chillers and pumps for process cooling applications from chillers and pumps functioning to control the environment and therefore creating dual temperature loops instead of a single temperature loop.

Tschudi et al. [2] recommended to develop an overall chilled water system efficiency (kW/t) by adding the efficiency of chilled water pumps, condenser water pumps and cooling towers. They represented a bench mark for different chilled water systems operating to cool typical cleanrooms. Based on their research the cooling system efficiency for water-cooled and air-cooled chilled water system, while setting the leaving chilled water temperature to 38, 40, and 42 °F for water-cooled system and 40, 42, 48 and 50 °F for air cooled systems could be benchmarked to slightly less than 0.7 kW/t for water-cooled system with a leaving water temperature of 42 up to near 1.1 kW/t for air-cooled system with a leaving water temperature of 48 °F. The best performance for both water-cooled and air-cooled systems happened where the leaving water temperature was set at 42 °F. Chilled water overall plant efficiency was slightly less than 0.7 kW/t for water-cooled (as noted earlier) and slightly less than 0.8 kW/t for the air-cooled system.

In addition to the cooling system, the heating system can have relatively significant effect on the central plant efficiency. The reason is the relatively large quantity of preheating requirement for the large quantity of outdoor air in heating season and also relatively large quantity of reheating requirement for dehumidification of all the supply air to the space during the cooling season. This as well calls for an efficient boiler, efficient hot water pumping system, and of course efficient hot water piping design to reduce the level of energy consumption when these functions are needed.

As it has been also noted in previous chapters for a thorough and in-depth discussion for learning different concepts associated with the cleanrooms and also designing the HVAC systems suitable for them refer to ASHRAE Application Handbook [1].

References

1. ASHRAEApplication Handbook. (2015). *ASHRAE handbook, HVAC applications*. Atlanta, GA: American Society of Heating, Refrigerating and Air conditioning Engineers.
2. Tschudi, B., Sartor, D., & Xu, T. (2001) An energy efficiency guide for use in cleanroom programming, supported by Northwest energy efficiency alliance and California energy commission.

Chapter 7
Commercial Kitchen and Dining Facilities

Abstract Design and operating commercial kitchens and associated dining areas are another one of the high energy consuming HVAC applications. Energy intensity (energy consumption per square foot) of kitchens arguably is one of the highest levels among other commercial applications specifically due to the relative small footprint of the commercial kitchens while the level of heat generation from cooking operation is extremely high. Based on a report published by US department of energy in 2006 food service facilities in the USA have an average energy intensity of 258,000 Btu/sq ft.

Keywords Commercial kitchen • Kitchen facilities • Hood • Grease • Wall mounted • Canopy • Scrubber • Volatile organic compounds • Demand control

Design and operating commercial kitchens and associated dining areas are another one of the high energy consuming HVAC applications. Energy intensity (energy consumption per square foot) of kitchens arguably is one of the highest levels among other commercial applications specifically due to the relative small footprint of the commercial kitchens while the level of heat generation from cooking operation is extremely high. Based on a report published by US department of energy in 2006 food service facilities in the USA have an average energy intensity of 258,000 Btu/sq ft. ASHRAE Standard Application [1] defines two main targets for the ventilation of the kitchen facilities. The first obvious purpose is keeping the kitchen environment comfortable and the second and even more important purpose is to provide a safe environment by properly removing the smoke, grease, and heat from the kitchen. Kitchen exhaust hood and its associated exhaust fan discharge a large volume of air in order to keep the results of cooking process out of the kitchen space. This large quantity of exhaust air needs to be replaced with proper amount of conditioned outside air. Conditioning this replaced air and delivering it to the kitchen just to be exhausted through the hood is the main source of high energy consumption in commercial kitchens.

Kitchen hoods are sub-categorized to Type I and Type II [1]. Type I hoods are used for removing grease, smoke or condensable vapor and are installed above appliances such as fryers and griddles. Type II hoods which are usually used for other applications such as dishwasher exhaust removal are generally designed for removing steam or heat. Hoods can be generalized to many different types. First type is the wall-mounted canopy style in which the wall behind the hood will act as back panel and therefore all the exhaust air will be drawn from front and sides of the appliance. This causes higher effectiveness in capturing effluent (combination of water vapor and particulate and vapor form organic material) that is developed by the cooking process. The other types are single or double island canopy in which the hood(s) are open on all sides and usually require more quantity of exhaust air to capture the developed effluent by the cooking process. There are also other types of hoods that are not as frequently used as the ones noted above such as proximity, eyebrow and pass-over hoods that each has its own characteristics. A well designed proximity type hood generally has the lowest demand for exhaust air flow capacity.

In practice different approaches have been used to control the effluent in discharge air as much as possible to protect the environment as well as people against diverse effects of kitchen hood exhaust to the outdoor. Strategies such as electrostatic precipitators, ultraviolet destruction, water mist and scrubbers, activated-carbon filters, incineration, and catalyst conversion. In electrostatic precipitators technology high voltage ionization is used to remove the particles, in ultraviolet destruction method ultraviolet light is used to chemically convert the grease into an inert substance, water mist and scrubbers trap particulates mechanically when effluent moves through water. Activated-carbon filters absorb VOC (volatile organic compounds), incineration technology uses high temperature oxidization to turn VOC and particulates to solid material, and catalyst conversion technology provides extra heat to decompose particulates and VOC [1].

An old solution for decreasing the large capacity of the replacing air was to introduce unconditioned or heated only air directly into the hood enclosure through a supply fan assembly which could have some sort of heating capability as well. This method generated a short circuit inside the hood enclosure that considerably reduced the capacity of the required conditioned air from the room to be exhausted through the hood. This approach is not generally pursued anymore due to both building codes requirement of 10 °F maximum allowable temperature difference between makeup air into the hood and kitchen air temperature and also researches that have showed in practice only 10–15 % of the short circuited air can be effectively implemented without allowing diffusion of the short circuited air into the kitchen space.

In most cases the dining spaces are located next to the kitchen area. Dining areas usually are designed to allow a high number of guests whom are required to be fed with considerable quantity of outdoor air. Therefore this is customary to allow the outside air that is provided for satisfying people's comfort in dining space to be transferred to the kitchen through transfer ducts or openings. This transferred air can be used as one major source of outside air delivery to the kitchen which will be discharged through the kitchen hood exhaust system (Fig. 7.1).

The general strategy for designing commercial kitchen and dining areas is to keep the whole building under positive pressure in order to prevent infiltration of unconditioned outdoor air into the building, while keeping a relative negative air pressure inside the kitchen itself to prevent the movement of kitchen odorous air from entering to the other parts of the building.

Most engineers feel comfortable with counting on 80–90 % of the replaced air being delivered through a 100 % dedicated outdoor air unit to the kitchen. This large quantity of pretreated air into the kitchen space then helps reducing the remaining required capacity of outside air to only 10–20 % of the replaced air by the hood, which should be provided by the transferred air or the outside air associated with another air handling unit that is designed to condition the kitchen in general.

If grease accumulates inside the duct or at fans casing or above the roof on the roof surface after leaving the fan it can act similar to fuel in case a fire breaks out in the building [1]. General building and fire protection codes and standards define the material, construction, and more importantly the volume of the air that should be exhausted through the different hoods. These exhausts are usually specified in cubic feet per minute per linear foot of the perimeter of the hoods or in a lesser degree some times in square feet of open area of the hoods. As the result of these regulations that pretty much dictate the required exhaust flow volume of the air through the hood there would be not much of a chance left for the designers to work around and reduce the exhaust air flow capacity for the hood. Therefore other energy saving and sustainability approaches should be considered to bring down the level of energy consumption in kitchen facilities.

Microprocessor based controls with sensors that control and vary the exhaust and supply fans capacities by using variable frequency drives based on presence of grease and temperature of the exhaust from the hood have significant effect of reducing the exhaust air volume and therefore the associated replaced conditioned air. These can be translated to considerable decrease in energy consumption for conditioning the air and fan power input as well.

ASHRAE Standard 90.1 [2] limits the quantity of the replaced air which is directly introduced into the hood enclosure to a maximum of 10 % while minimizing the quantity of the conditioned air introduced to the kitchen to not greater than:"(1) supply air required for space heating and cooling and (2) hood exhaust flow minus transfer air from adjacent spaces which is not counted for by other exhaust systems" [2]. AHSRAE Standard 90.1 [2] based on the type of duty of the equipment under the hood (light, medium, heavy, and extra heavy) specifies the minimum net exhaust flow for conditions that the total exhaust from the kitchen hood is above 5000 cubic feet per minute. If a kitchen has a total hood exhaust capacity of larger than 5000 cubic feet per minute it also should be equipped with one of the following options:"(1) At least 50 % of the exhaust air should be introduced to the kitchen via transferred air that otherwise was exhausted, (2) demand control on at least 75 % of the exhaust air, or (3) heat recovery device with at least 40 % sensible heat recovery effectiveness on at least 50 % of the exhaust air capacity" [2].

ASHRAE Application Handbook [1] in addition to the exhaust air quantity names a few other major factors that have a great influence on energy consumption level of commercial kitchens, such as climate, system operating hours, static pressure and fan efficiencies, and radiant load of the appliances located under hood.

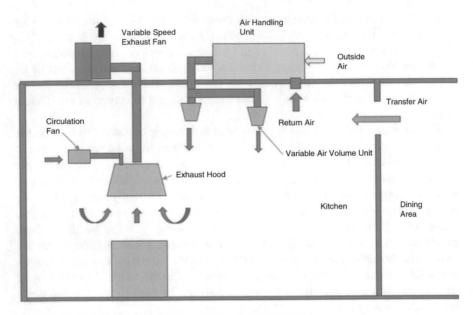

Fig. 7.1 Typical kitchen variable air volume exhaust strategy

Some of the most recommended strategies for decreasing the level of energy consumption in commercial kitchens are selecting and specifying hoods that require lowest amount of exhaust air flow, installing side panels, maximizing level of transferred air, demand control ventilation, heat recovery from hood exhaust where it is possible, optimizing the air set points, air side economizer controls, etc.

A demand controlled kitchen ventilation is basically an approach of modulating down the required airflow exhaust from the kitchen hood in part load condition. The difference between the demand controlled kitchen ventilation and the usual demand control ventilation which is the common way of energy saving in office buildings or similar applications is that the former one is done by modulating the air flow based on the cooking operations while the latter one is based on level of CO_2 presented in the space. Optimizing the air set point comes from the code requirement of maximum allowable temperature difference of 10 °F between the room and the introduced air to the hood. Since the temperature for maintaining the comfort level in the kitchen (~85) generally is much higher than other applications (~75 °F) this provides the opportunity for optimizing the set point to a level in which considerable energy saving can be resulted.

If we look a little more in depth into each of these guides presented for energy efficiency in data centers, health-care centers, laboratories, cleanrooms, and commercial kitchens, soon we recognize that next to selecting units with higher efficiency, the most important factor in achieving this goal is a well-written and well-implemented control system. Controls that have a major role in any other application energy saving have even more importance when the level of energy consumption in an application is so high. Selecting the best possible systems and writing a very effective control sequence to get the systems work together requires very effective and well-designed decisions made by the designers. The designer's proper decision-making becomes a vital part of having energy-efficient systems. In the following chapters of this book I attempt to introduce some of the most advanced decision-making techniques which would be tremendously useful if they become a part of the tools that are in a daily basis used by the engineers and designers.

References

1. American Society of Heating, Refrigerating and Air Conditioning Engineers. (2015). *ASHRAE handbook application*. Atlanta, GA: American Society of Heating, Refrigerating and Air Conditioning Engineers.
2. American Society of Heating, Refrigerating and Air Conditioning Engineers. (2013). *ASHRAE 90.1 Standard. Energy standard for buildings except low-rise residential buildings*. Atlanta, GA: American Society of Heating, Refrigerating and Air Conditioning Engineers.

Part III
Advanced Decision Making Strategies

Chapter 8
Introduction

Abstract The Webster dictionary definition of decision is "a choice that you make about something after thinking about it." There is a large amount of research done on specifying number of decisions that each person makes during his lifetime. Numbers are widely different as it is hard to draw a clear line to distinguish between conscious and unconscious decisions, and therefore, these numbers are varied between a few and extremely high for a typical person in a regular daily basis. Whatever the real number of our decisions in a day is, it is obvious that some of them are of a very minimal importance and some are more significant that can carry substantial consequences, either good or bad.

Keywords Decision • Prescriptive • Normal • Descriptive • Positive • Decision-making under uncertainty • Multi-criteria decision-making • Decision support system • Epistemic • Aleatory • Utility function • Value function

The Webster dictionary definition of decision is "a choice that you make about something after thinking about it." There is a large amount of research done on specifying number of decisions that each person makes during his lifetime. Numbers are widely different as it is hard to draw a clear line to distinguish between conscious and unconscious decisions, and therefore, these numbers are varied between a few and extremely high for a typical person in a regular daily basis. Whatever the real number of our decisions in a day is, it is obvious that some of them are of a very minimal importance and some are more significant that can carry substantial consequences, either good or bad. How much value would a person have whom he offers us tools and methods to equip us with required knowledge to make better decisions? Probably it is priceless, since correctness of the decisions or their failures makes or breaks the success of individuals, engineers, companies and even nations. In 1738 and then later in 1947 Daniel Bernoulli with "Specimen theoriae novae de mensurasortis (Exposition of a New Theory on the Measurement of risk)," and von Neumann and Morgenstern with (von Neumann–Morgenstern utility theorem) offered us such tools, by defining vital concepts of value function and utility function respectively. These tools have become the corner stones of the modern decision-making approaches.

Every engineer and architect like most of the other people is continuously faced with different problems that in order to be solved in a proper manner are required to

© Springer International Publishing Switzerland 2016   65
J. Khazaii, *Advanced Decision Making for HVAC Engineers*,
DOI 10.1007/978-3-319-33328-1_8

be evaluated and proper decisions to be made about them. The general field that discusses the decision-making process and its related topics is called decision theory. Two different branches under the umbrella of decision theory are defined to be normative (prescriptive) and positive (descriptive) methods. Normative or prescriptive type is part of the decision theory which is concerned with how people should make decision. The main requirement of this category is that it assumes a capable, well-equipped, well-informed rational decision maker. On the other hand, positive or descriptive decision theory is just concerned with observation and explanation of how people make decision. The assumption here is that people make decisions based on some accepted rules or framework. Therefore, it could be interpreted that the way that people "make" their decisions is subject of the descriptive decision-making which is more relevant in psychology and behavioral type of science and the way that they "should make" their decisions is subject of the normative decision-making which is more relevant in economics and engineering type of sciences. Our focus in this book is completely directed towards normative decision-making which is assumed to be made by a rational, well equipped and well informed decision maker (engineer).

Generally applying the normative method for making proper rational decisions under such circumstances is called "Decision Analysis." Different scientists have categorized decision analysis into different subcategories. One of the mostly accepted of these definitions is specified by Zhou [1, 4]. He categorizes decision analysis into three main subsets of (1) decision-making under uncertainty, (2) multi-criteria decision-making (MCDM), and (3) decision support system (DSS). To name samples of these categories we can point out the use of utility functions and decision trees, multi-objective optimization, and intelligent decision support system in each category respectively. In the future chapters we will discuss these categories and associated methods in some more depth, but here we just suffice to a short and introductory definition of each of these three main categories which are the subject of the majority part of the rest of this book.

Decision-making under uncertainty is the process of making decisions about actions with unknown inputs and outcomes. Uncertainty is divided to aleatory uncertainty and uncertainty due to lack of knowledge (epistemic). The level of certainty in an aleatory uncertain problem does not improve with unveiling additional and new data to the system, because the data is in its nature random and therefore sometimes is one quantity and other times is another. It means even if we throw a fair coin 100 times, the next outcome is still a 50-50 chance for showing head or tail, or even if we gather years and years of the past weather data, we still will not be capable of exactly predicting the next year weather condition. To the contrary introducing every additional piece of data improves the strength of an epistemic uncertain problem. It means for example a table has an exact length and weight and with more measurement we can get more information that takes us closer and closer to the exact length or weight of that table. A common way of permeation of uncertainty to our models is through either uncertain input or the uncertain structure of the model itself, where the latter one tends to deal more with the underlying mathematics of the model used for simulation, while the former one which we usually get more involved with (input uncertainty) most of the times deals with the lack or shortage of information about what we use as input to the simulation model. This input is in some degree different from what is actually used in the real world when they build the actual object and in our case actual building. It should be noted that the input uncertainty can be either epistemic or aleatory kind, while the model uncertainty is usually categorized as epistemic type. Some times when we cannot objectively quantify the probability of an input data, we use a subjective probability method to do so. While an objective probability data can be derived from real world data and collectable empirical information, a subjective probability data is usually assessed based on the level of confidence and belief of the modeler. Bayesian rules are usually being utilized in these cases. Most of the times when modelers model an uncertainty problem, they try to avoid including aleatory or subjective uncertainty in their model because if its assigned probability is not supported with properly structured methods and evidences, the model almost certainly will produce unreal results. Most modelers usually try to use epistemic uncertainty related to the objective data as the base of the uncertainty analysis to keep the output closer to the reality, and if they are forced to include subjective data into their simulation, they may use that as the scenario under which the simulation takes place. Most of the input uncertainty that is generally used in building simulation, is drawn from the allowable tolerances for material character-istics and efficiencies, regulated by testing agencies such as Air Conditioning, Heating, Refrigerating Institute (AHRI).

Understanding and calculating uncertainty is only one part of the equation in decision-making under uncertainty, but to be able to make decision based on the calculated uncertainty, we need a framework to use as the basis for our decision-making. Shultze et al. [2] have presented an informative history about the evolution of this framework or the utility function. Until early 1700s and before presentation of St Petersburg Paradox the conventional belief of scientific community was that individuals would select their actions with a target of "Maximizing wealth." Nickolaus Bernoulli presented the St Petersburg paradox and questioned the valid-ity of this belief. Daniel Bernoulli in mid 1700s partly resolved this paradox with presenting concepts of "Value Function" and "Diminishing Marginal Value of Wealth," but that still was not sufficient to completely answer why people are not indifferent between a sure payment and a payment with the same expected value derived from the presented value function. In mid 1900s von Neumann and Morgenstern introduced the concept of "Utility Function" or "Attitude of the people toward Risk and Wealth" and also defined the "Expected Utility" which is the expected value of the utility function, which finally was able to resolve the St Petersburg paradox completely after more than 200 years. A few years later, Savage showed the subjective expected utility functions also can be used if proper effort is done to assign the correct percentages to these values. To implement a subjective probability analysis one of the most common methods is utilizing Bayesian rules. "In settings with uncertainty, where all relevant probabilities are objective and known, we call an agent NM (von Neumann and Morgenstern) rational if he acts as if he is maximizing a NM expected utility function. What if the probabilities are not given? We call an agent Bayesian rational (or say that he has subjective expected utility preferences) if: (1) in settings with uncertainty, he forms beliefs describing the probabilities of all relevant events. (2) When making decisions, he acts to maximize his expected utility given his beliefs. (3) After receiving new information, he updates his beliefs by taking conditional probabilities whenever possible" [3].

Finally it should be noted here that there are multiple methods for choosing an optimum utility function, to name a few we can call maximum of the minimums, or weighted utility method.

Moving to the second group of decision analysis tools, we can define a multi-criteria decision-making process as a method which helps the decision makers by giving them access to different tools which by using them the decision maker will be capable of solving complex problems while targeting to improve multiple criteria in the same problem. There are different approaches recognized in multi-criteria decision-making such as simple multi-attribute rating technique, analytical hierarchy process, and analytical network process. Sometimes the designer's interest is not just to make a very good decision but also to optimize his decisions. It means he wants to get the best possible answer(s). The techniques for optimization are either deterministic or stochastic (uncertain). A deterministic optimization method in general is either single-objective optimization (SOO) or multi-objective optimization (MOO). A stochastic optimization is called a robust optimization or optimization under uncertainty [4].

Savic [5] explains: "The main goal of single-objective optimization is to find the "best" solution, which corresponds to the minimum or maximum value of a single objective function that lumps all different objectives into one. This type of optimization is useful as a tool which should provide decision makers with insights into the nature of the problem, but usually cannot provide a set of alternative solutions that trade different objectives against each other. On the contrary, in a multi-objective optimization with conflicting objectives, there is no single optimal solution. The interaction among different objectives gives rise to a set of compromised solutions, largely known as the trade-off, non-dominated, non-inferior or Pareto-optimal solutions."

Well-known single-objective optimization methods are Newton–Raphson, Hooke–Jeeves, and genetic algorithms [4]. On the other hand, multi-objective optimization can be divided to two sub-categories of Pareto base and non-Pareto base multi-objective optimization. The most famous non-Pareto base optimization arguably is the game theory based on work of von Neumann which later was infused into the economics by John Nash equilibrium. The other multi-objective optimization method is Pareto base optimization which can be subdivided to (1) - decision-making with a priori Pareto optimization (one point in Pareto front), (2) decision-making with a posteriori Pareto optimization (Pareto front), and (3) - decision-making with progressive Pareto optimization [4]. Under multi-objective optimization method, the target of Pareto optimization is to bring the system to a state that it is not possible to improve one individual (element) to a better status without making the other individuals (elements) worse off. In other words, there is no other design possible such that the presented solution is at least equal in all characteristics and is better in at least one characteristic than other solutions.

Furthermore the difference between a priori and a posteriori Pareto optimization methods is that in a priori method sufficient preference information or relative importance (weighted preferences) should be assigned before the optimization process (a deductive approach) starts, but in a posteriori method it is required that all the Pareto optimal solutions to be generated first and then the designer should choose from this frontier (an inductive method). Well-known types of a priori method are utility function, lexicographic, and goal programming methods. (A no preference method is basically similar to a priori method but does not require weights of the preferences to be included in optimization). A posteriori method can be subdivided into two main categories of mathematical programming and evolutionary algorithms. Among mathematical programming methods we can name normal boundary intersection (NBI), modified normal boundary intersection (NBIm), normal constant (NC), etc. Among evolutionary algorithm methods we can name multi-objective genetic algorithm (MOGA), non-dominated sorting genetic algorithm-II (NSGA-II), strength Pareto evolutionary algorithm 2 (SPEA-2), particle swarm optimization, etc [4].

In a progressive or interactive method the modeler in each step of optimization includes his/her own preferences such as trade-off information and reference points and repeats the optimization one after another until coming up with the final Pareto optimal solution.

Hybrid methods can be also created by combining algorithms from different fields of optimization.

Finally common methods that have been used in robust optimization are Kriging models and neural network methods [4].

It should be noted that in all of these methods the expert judgement can play a major role depending on the type of the problem and accessibility of the data for design problem. In some literature the expert judgement by itself creates a subcategory of the multi criteria decision-making.

The third category in Decision Analysis is Decision Support decision support system (DSS)Systems which is defined as a computer-based information system that can help individuals, companies and businesses to make proper decisions for their activities and in dealing with their problems. Sprague [6] defines decision support system characteristics as, aiming at the less well structured, underspecified problems, combining the use of model or analytic techniques with traditional data access and retrieval functions, focusing on those features which make them easy to use and emphasizing on flexibility and adaptability to accommodate changes in the environment and decision-making approach of the user. Holsapple and Whinston [7] classified decision support systems into (1) text-oriented, (2) database-oriented, (3) spreadsheet-oriented, (4) solver-oriented, and (6) rule-oriented systems. Power [8] also subdivides decision support system to (1) data driven, (2) model driven, (3) knowledge driven, (4) document driven, and (5) communication driven types.

Different literatures state that decision-making can be as easy as a selection among clearly available options, or in a more advanced type be also included the process of gathering the possible alternatives as well, or in a very high level on top of all of these be included of searching for opportunities for decision-making in a non-reactive fashion. Samples for each level could be turning the air conditioning system on or off, seeking alternative methods for air conditioning such as using natural ventilation where it is possible, and optimizing the operation between forced and natural ventilation to maximize energy saving without compromising the level of comfort. By digging inside these definitions it could be implied that the target of a well-developed decision support system should be the broadest final approach. A properly designed decision support systems not only should help in finding the best solutions for a given question, they also should be capable of helping in unexpected emergence of opportunities and its solution as well.

"Decision Support Systems are especially valuable in situations in which the amount of available information is prohibitive for the intuition of an unaided human decision maker and in which precision and optimality are of importance. Decision support systems can aid human cognitive deficiencies by integrating various sources of information, providing intelligent access to relevant knowledge, and aiding the process of structuring decisions" [9].

One of the tools in this category which has gained some momentum in order to not only describe the technical aspects of the problems, but also has made possible a big advancement in understanding and studying the social behavior of people, is agent based modeling. Agent based modeling has common structure with the gaming theory and allows the modeler to design agents which are capable of making decision and let them to interact with each other and with the environment in order to discover possible trends and emerging situations. An intelligent decision support system uses intelligent agents which interact with each other and the environment, learn and use their learnings for achieving goals. Russell and Norvig [10] categorizes intelligent agents into (1) simple reflex, (2) model-based reflex, (3) goal-based agents, (4) utility-based, and (5) learning agents.

In the next few chapters I try to introduce some examples of each one of the three major categories of decision analysis formatted to fit the applications of HVAC, architectural, and energy engineering. My main concern is not to couple multiple software together and present results from the simulation, but is to cut through the logics of each presented method and explain the procedure that a software or a person has to undertake in order to advance the processes, models, and simulations. I invite the interested researchers to try to use different software to write their own decision-making tools based on the patterns that are presented and couple them with energy modeling software such as "Energy Plus" where it is desired to develop automated and faster results.

References

1. Zhou, P., Ang, B. W., & Poh, K. L. (2006, November). Decision analysis in energy and environmental modeling: An update. *Energy, 31*(14), 2604–2622. doi:10.1016/j.energy.2005.10.023.
2. Schultz, M. T., Mitchell, K. N., Harper, B. K., & Bridges, T. S. (2010) *Decision making under uncertainty*. ERDC TR-10-12. Vicksburg, MS: U.S. Army Engineer Research and Development Center.
3. Sandholm, W. H. (2014). Lecture notes on game theory. http://citeseerx.ist.psu.edu/viewdoc/download?doi=10.1.1.676.3870&rep=repi&type=pdf
4. Hopfe, J. C. (2009). Uncertainty and sensitivity analysis in building performance simulation for decision support and design optimization; PhD Thesis.
5. Savic, D. (2002). Single-objective vs. Multi-objective optimization for integrated decision support. Integrated Assessment and Decision. *Proceedings of the First Biennial Meeting of the International Environmental Modelling and Software Society*.
6. Sprague, R. (1980). A framework for the development of decision support systems. *MIS Quarterly, 4*(4), 1–25.
7. Holsapple, C. W., & Whiston, A. B. (1996). *Decision support systems: A knowledge-based approach*. St. Paul, MN: West Publishing. ISBN 0-324-03578-0.
8. Power, D. J. (2002). *Decision support systems, concepts and resources for managers*. https://books.google.com/books?id=9NA6QMcte3cC&pg=PA2&lpg=PA2&dq=types+of++design+decision+support+system&source=bl&ots=DNpumDRxEd&sig=yNzu_TIh3ScBXchrgJri E69EmdA&hl=en&sa=X&ei=tKGMVY2AAZXboASitZfoCA&ved=0CDwQ6AEwAjgK#v=onepage&q=types%20of%20%20design%20decision%20support%20system&f=false
9. Druzdzel, M. J., & Flynn, R. R. (2002). Decision systems Laboratory, school of information sciences and intelligent Systems Program, University of Pittsburgh, PA.
10. Russell, S. J., & Norvig, P. (2003). *Artificial intelligence: A modern approach* (2nd ed.). Upper Saddle River, NJ: Prentice Hall. ISBN 0-13-790395-2.

Chapter 9
Analytical Hierarchy Process (AHP)

Abstract Analytical hierarchy process is a method in multi-criteria decision-making field that helps the designer to select proper solution in complex multi-objective problems. The origin of the method goes back to 1970–1980s when Professor Thomas L. Saaty originated and developed this method. In his words "The Analytic Hierarchy Process (AHP) is a theory of measurement through pairwise comparisons and relies on the judgements of experts to derive priority scales. It is these scales that measure intangibles in relative terms. The comparisons are made using a scale of absolute judgements that represents, how much more, one element dominates another with respect to a given attribute."

Keywords Analytical hierarchy process • Multi-criteria decision making • Main criteria matrix • Ranked criteria • Normalized criteria • Relative priority • Consistency • Credibility • Eigen value • Eigen vector

Engineering problems in general and building-related design issues in particular have become further and further more complicated. This complicated process requires the integration and cooperation of multi-discipline engineers, architect, owner, equipment manufacturers, etc. Each agent has its own expertise and therefore is responsible and enthusiastic for his or her part of the design to meet all the trade-required objectives and criteria. It is obvious that some of these objectives are conflicting such as saving energy and increasing the level of comfort of the occupants. In the meantime some of these objectives are more important than the other objectives, such as the desire of the architect to make a higher impact by the form of the building compared to the desire of the mechanical engineers for having larger service space for the equipment above the ceilings. Lack of existence of a well-understood strategy from day 1, among the main players in this integrated exercise, could easily lead to a product that not only is not capable to satisfy all the player's expectations properly, but also can almost guaranty the spending of the money and resources to advance targets that may not be the most common agreed-upon team priorities. Of course in the past decade LEED (Leadership in Environmental Engineering Design) scoring system or similar methods have attempted to create some rules for the playground by getting the different players to agree upon some general directions in advance, like commitment of mechanical engineer to avoid using improper refrigerants in the system and working towards some

© Springer International Publishing Switzerland 2016
J. Khazaii, *Advanced Decision Making for HVAC Engineers*,
DOI 10.1007/978-3-319-33328-1_9

minimum energy efficiency in the building or commitment of the plumbing engineer to design with provisions to save some minimum level of water usage. But these methods generally fall short of creating a framework that can be used as a measure for combining the desire of the designers, owners, etc. for selecting different alternatives for the design problem, specifically when these desired objectives cannot be expressed with the same dimension. It can be argued that, e.g., saving some percentage of overall energy of the building really should not be equated with having one accredited professional in the design team or having storage capacity for bicycles as it is valued similarly for acquiring credit points in LEED scoring system.

Analytical hierarchy process is a method in multi-criteria decision-making field that helps the designer to select proper solution in complex multi-objective problems. The origin of the method goes back to 1970–1980s when professor Thomas L. Saaty originated and developed this method. In his words [1] "The Analytic Hierarchy Process (AHP) is a theory of measurement through pairwise comparisons and relies on the judgements of experts to derive priority scales. It is these scales that measure intangibles in relative terms. The comparisons are made using a scale of absolute judgements that represents, how much more, one element dominates another with respect to a given attribute."

In developing this method Saaty has created a fundamental scale of absolute numbers including 1/9, 1/7, 1/5, 1/3, 1, 3, 5, 7, and 9 (and 2, 4, 6, 8, 1/2, 1/4, 1/6, and 1/8 when partial tie-breakers are needed) and has assigned values of equal importance (1), moderate importance (3), strong importance (5), very strong importance (7), and extreme importance (9) to these numbers. Multiple experts with deep knowledge and also people (in building design case, owner, government agents, etc.) who have interest in solving a specific problem will use this ranking to provide a judgement matrix showing the relative importance of what they believe are the main criteria components related to the problem in hand, and as a result provide a normalized priority scale for the main criteria parameters. Similar approach will be done to provide priority relation for sub-criteria (if exist) of the main criteria. Finally a pairwise comparison based on these developed priorities will be done for the solution alternatives. Combination of these priority relations among the main criteria and alternatives will provide a final normalized listing that would help the designer to select the best solution for his problem. For a detail guide for how to use this method see Saaty [1] where he explains how to quantify the relative consumption of drinks in the USA by using analytical hierarchy process. In general this process requires a detail pairwise comparison between each two of the problem criteria and some basic knowledge of matrix functions, and in return provides a relatively strong quantified comparison among the problem solution alternatives.

Triantaphyllou and Mann [2] has pointed out one of the most important advantages of analytical hierarchy process as follows: "In a simple MCDM situation, all the criteria are expressed in terms of the same unit (e.g., dollars). However, in many real life MCDM problems different criteria may be expressed in different dimensions. Examples of such dimensions include dollar figures, weight, time, political impact, environmental impact, etc. It is this issue of multiple dimensions which makes the typical MCDM problem to be a complex one and the AHP, or its variants, may offer a great assistance in solving this type of problems." In the meantime as a probable weakness they also strongly suggest that when some of the alternatives end up to be very close to each other decision maker should be very cautious and need to consider more criteria to help creating obvious gaps between alternatives.

As a practical application of the method, Sloane et al. [3] represented a study that helps to show the proper weighting for competing criteria for capital investment in modern hospitals. In another study Kousalya et al. [4] applied analytical hierarchy process for selecting a student from an engineering college who is eligible for all-round excellence award.

In engineering field the designers are always facing multiple decision-making problems. For example in the field of building design the mechanical, electrical, and architectural engineering firms are always in need to select among different building materials, different equipment, and different design configurations. They also are in need to integrate the interests of the owner of the building, and consider the well-being of the public and many other important factors into their design as well. This creates a situation that even most experienced engineers cannot simply select the most suitable alternative for satisfying all the parties that benefits from the design to its maximum possible. Usually in this condition designers consider some compromised design options which are built based on their best judgement. This compromise or decision making is a necessity for work to be advanced, but unfortunately it might occur without real scientific base which should be considering all the factors, and their actual weights, and is usually done only based on some best guess assumptions of the seasoned designers or based on the owner's money restrictions. Using scientific based exercises such as one available with analytical hierarchy process if it is implemented can create a breakthrough in this compromised decision making for the players of the building design industry in order to have an optimized solution which satisfies the most possible requirements presented by each player.

As I pointed out earlier, in analytical hierarchy process method the analyst first has to decompose the complex problem into smaller and more comprehensible hierarchy of problems. These smaller elements should be related to the problem aspects and shall be compared one by one against each other. The final decision can be made based on the normalized relative value of the alternatives, which will be calculated using the overall normalized relevance factor of the problem aspects. Both use of eigenvectors or higher powers of a decision matrix have been recommended in different literatures as possible ways of generating the normalized preferred relationship between parameters of the problem. By using these methods one can select one alternative from a pool of alternatives when there is multiple decision criteria and different stakeholders have different preferences. Let's dive into the process by means of solving a hypothetical HVAC-related problem.

In order to set up our exercise for analytical hierarchy process example let's go back and use the same multiple story office building in Atlanta that we discussed previously. Here our decision to make is what type of HVAC system is more suitable for this building based on some specific criteria. Even though the process will be the same for multiple systems, but in the benefit of simplicity, let's narrow our problem further to find out which one of these two systems, (1) water-cooled unitary system with electric stripes or (2) air-cooled unitary system with electric stripes, would be the most desirable choice for this building, given the specific criteria of our choice.

The first step in setting up an analytical hierarchy process decision making is to depict a matrix that defines the main criteria of the problem in hand. In this matrix we can assign the relative values to the main criteria of the problem. This step requires input from a group of the experts and also parties which hold interest in the building. The traditional way of creating this matrix as was stated earlier is based on a spectrum of numbers between 1/9 and 9 stating the least desirables and most desirable ends of each criteria as they are compared one by one and by a group of experts and interested parties in the problem. Here let's adopt another way of developing this matrix based on what has been shown in LEED scoring system criteria and assume this revised take of the LEED criteria as our basis for developing the analytical hierarchy process main criteria matrix.

LEED version 2.2 defines seven main criteria as the possible fields that can contribute to the sustainability of the building. These seven criteria are (1) sustainable sites (24 points), (2) water efficiency (11 points), (3) energy and atmosphere (33 points), (4) material and resources (13 points), (5) indoor air quality (19 points), (6) innovation and design process (6 points), and (7) regional priority credits (4 points), for a total of 110 available credit points. Let's for simplification of our work disregard the last two categories which are equivalent for ten credit points and make our main matrix based on the first five criteria and associated credit points of each with a total of 100 credit points. Therefore assume we have gotten together the interested parties in our building and they have voted their most desired (important) factors in design of this building and its relevance as follows:

The values in this matrix have been calculated based on the relative values of the credit points of the rows to the columns, e.g., 2.18 specified in site row and water efficiency column is calculated by dividing 24 (total credits assigned to the site conditions) by 11 (total credits assigned to water efficiency criteria) and 0.46 specified in water efficiency row and site column is the result of 11 divided by 24. Even though it is possible to use these values to develop the normalized preference value of the main matrix also, but in order to stay closer to the structure of the traditional version of analytical hierarchy process, we further revise the numbers in the matrix above to either one of 1/9 to 9 values. By substituting the below table (Table 9.1) with the traditional matrix format defined by AHP and some approximation we will have a table similar to Table 9.2. The process of this function is to compare each two pairs separately and decide if its relationship as shown in Table 9.1 is close to which one of the values in AHP 1/9 to 9 relationship criteria. By performing this exercise each place in the matrix below (Table 9.2) will be filled with one of 1, 0.33, 3, 0.2, 5, 0.14, 7, 0.11, or 9 values. As I mentioned before the Saaty's method allows for intermediate values such as 2 and 4. But here for the benefit of simplification of calculations we will stay with the original 1/9 to 9 values.

Table 9.1 Main criteria matrix

	Site	Water efficiency	Energy and atmosphere	Material use	Indoor air quality
Site	1.00	2.18	1.26	1.85	0.73
Water efficiency	0.46	1.00	0.58	0.85	0.33
Energy and atmosphere	0.79	1.73	1.00	1.46	0.58
Material use	0.54	1.18	0.68	1.00	0.39
Indoor air quality	1.38	3.00	1.74	2.54	1.00
	4.17	9.09	5.26	7.69	3.03

Table 9.2 Main criteria ranked in 0.11–9

	Site	Water efficiency	Energy and atmosphere	Material use	Indoor air quality
Site	1.00	5.00	1.00	3.00	0.33
Water efficiency	0.20	1.00	0.33	1.00	0.20
Energy and atmosphere	1.00	3.00	1.00	3.00	1.00
Material use	0.33	0.33	0.33	1.00	0.20
Indoor air quality	3.00	5.00	1.00	5.00	1.00
	5.53	14.33	3.66	13.00	2.73

Table 9.3 Main criteria normalized matrix {C}

Normalized	Site	Water efficiency	Energy and atmosphere	Material use	Indoor air quality
Site	0.18	0.35	0.27	0.23	0.12
Water efficiency	0.04	0.07	0.09	0.08	0.07
Energy and atmosphere	0.18	0.21	0.27	0.23	0.37
Material use	0.06	0.02	0.09	0.08	0.07
Indoor air quality	0.54	0.35	0.27	0.38	0.37
	1.00	1.00	1.00	1.00	1.00

The next step is normalizing the above matrix (Table 9.2). The process of normalizing the matrix is to divide all members of each column one by one by the column total sum value in order to create a column that the total sum value of its members becomes equal to 1. Therefore the values in the first column of the normalized matrix will be calculated as 1/5.53 for first row, 0.2/5.53 for second row, etc. After completing this process for all the columns of the matrix {C} shown in Table 9.3, we will have the matrix which is the normalized version of matrix shown in Table 9.2. The eigenvector can be developed based on a higher power value of each row (a simplified method represented by Saaty as nth root of multiplication of the row's values, or traditional eigenvector calculation) and

Table 9.4 Relative priority
of the main criteria

{W}
0.22
0.07
0.24
0.26
0.37
1

represents the relative priority of the main criteria as it is shown above in matrix {w} which is based on a simplified nth root approach (Table 9.4).

This matrix {w} will be utilized at the end of the process as a factor in calculating the problem alternative relative values. But before going any further we should check the creditibility of the values filled in the main matrix. This is done to make sure that there is no irregularity in assigning relative values to the elements of criteria. By that Saaty meant that we want to make sure that the input to the main matrix is homogenous, and erroneous statements such as A is favorable to B and B is favorablc to C but A is not favorable to C do not exist.

Following the guidance represented by Saaty, and in order to perform a consistency check on values of the matrix {C} we need to define {Ws} as a matrix product of matrix {C} and {W}, and a consistency matrix {Consistency} as a matrix product of {Ws} and {1/W}.

The consistency matrix produces an array of values of Lambda (λ) which the largest of them is called maximum Lambda (λ_m) which is also called the eigenvalue of the matrix {C}. At this point we can calculate the Consistency Index (CI) which is calculated based on the following equation:

$$Cl = (\lambda_m - \text{Order of matrix } C)/(\text{Order of matrix } C - 1) \qquad (9.1)$$

The final step is to calculate the credibility ratio (CR) based on the following equation:

$$CR = CI/RI \qquad (9.2)$$

For a detail description for calculating credibility ratio see ref. [5].

RI is known as Random Consistency Index. General values for RI are given by Saaty. These values are shown in Table 9.5 below and can be used in different problems to help calculating the credibility ratio.

This process should be followed for every matrix developed in this system. For any matrix developed here the consistency ratio should be smaller than 0.1, or the assigned values in the matrix should be re-evaluated for possible inconsistency among the matrix element values.

Table 9.5 Random consistency index

n	1	2	3	4	5	6	7	8	9	10
RI	0	0	0.58	0.9	1.12	1.24	1.32	1.41	1.45	1.49

Table 9.6 Comfort matrix

	Outdoor air	Acoustical comfort	Thermal comfort	Daylight and view	Pollution
Outdoor air	1.00	5.00	3.00	3.00	1.00
Acoustical comfort	0.20	1.00	0.20	0.20	0.20
Thermal comfort	0.33	5.00	1.00	1.00	0.33
Daylight and view	0.33	5.00	1.00	1.00	0.33
Pollution	1.00	5.00	3.00	3.00	1.00
	2.86	21.00	8.20	8.20	2.86

Table 9.7 Comfort matrix normalized

Normalized	Outdoor air	Acoustical comfort	Thermal comfort	Daylight and view	Pollution
Outdoor air	0.35	0.24	0.37	0.37	0.35
Acoustical comfort	0.07	0.05	0.02	0.02	0.07
Thermal comfort	0.12	0.24	0.12	0.12	0.12
Daylight and view	0.12	0.24	0.12	0.12	0.12
Pollution	0.35	0.24	0.37	0.37	0.35
	1.00	1.00	1.00	1.00	1.00

As it can be shown the consistency ratio for matrix $\{C\}$ in our example is fulfilling this criteria easily.

After developing the relative values of the main criteria, the next step is to develop similar matrices and values for subcategories. In our work let's assume that there is no subcategory for site, water efficiency, and material categories, while energy and atmosphere can be subcategorized to two sections of energy efficiency and renewable energy and indoor air quality can be subcategorized to five sections of outdoor air, acoustical comfort, thermal comfort, daylight and view, and pollution. Tables (9.6, 9.7, 9.8, 9.9, 9.10, and 9.11) show our process of developing relative values for these two subcategories, which is basically very similar to what we did for the main matrix.

Table 9.8 Comfort relative
priority

{W}
0.33
0.04
0.14
0.14
0.33
1

Table 9.9 Energy matrix

	Energy performance	Renewable energy
Energy performance	1.00	7.00
Renewable energy	0.14	1.00
	1.14	8.00

Table 9.10 Energy matrix normalized

Normalized	Energy performance	Renewable energy
Energy performance	0.88	0.88
Renewable energy	0.13	0.13
	1.00	1.00

Table 9.11 Energy relative
priority

{W2}
0.88
0.12
1.00

Each of these matrices has proven to pass the consistency ratio requirement as defined earlier. By completing this step we have reached to the point that we can give a proper relative value to each of the main and subcategory criteria. It should be reemphasized here that these values have been developed based on the input from the interested parties' preferences and expert judgments. If we combine the information that we have gathered up to this point about the overall criteria in one table we can develop something similar to Table 9.12 below where multiplication of the values in each column can be the representative of the relative value (weight) of each parameter in the overall skim of the problem question. This means that for any building evaluated under this criteria, the site overall value is equivalent of (0.22×1), water efficiency overall value is equivalent to (0.07×1), energy efficiency overall value is equivalent to (0.24×0.88), etc.

After developing the above framework, now it is time for comparing our design alternatives according to this framework, and select which one of these alternatives would perform better under these general frameworks.

Table 9.12 Overall relative priority of the main and subcriteria

Site	Water efficiency	Energy and atmosphere		Material use	Indoor quality				
		Energy efficiency	Renewable energy		Outside air	Acoustical comfort	Thermal comfort	Daylight and view	Pollution
0.22	0.07	0.24	0.24	0.06	0.37	0.37	0.37	0.37	0.37
1	1	0.88	0.12	1	0.33	0.04	0.14	0.14	0.33

The above ten criteria (site, water efficiency, energy efficiency, renewable energy, material use, outside air, acoustical comfort, thermal comfort, daylight and view, and pollution) should be compared one by one against each other for all (both here) of the alternatives. This requires using proper techniques in order to compare these parameters in our selected alternatives. Methods can be either based on specific calculations or the expert input or literature research.

Let's remember our goal again which was to find out which one of these two systems, (1) water-cooled unitary system with electric stripes or (2) air-cooled unitary system with electric stripes, would be the most desirable choice for our example multi-story office building, given the criteria defined by the above ten parameters. In order to perform this evaluations let's separate the main criteria to two groups. Let's the first group to be site, renewable energy, material use, outside air, acoustical comfort, thermal comfort, and daylight and view which are either very similar for both alternatives or an expert can judge the advantage of one alternative over the other. The second group or water efficiency, energy efficiency, and pollution are those which can be calculated via performing two sets of energy modeling for our two alternatives.

From experience or expert input one can reach a conclusion that the effects of these two systems on site, renewable energy, thermal comfort, and daylight and view are almost identical and normalized value of 0.5 can be assigned to either of the two alternatives (of course more detail calculation can also be done in each case to calculate relative weights that are closer to the real values). Also from experience or expert input we can conclude that since the main air-handling units in water-cooled unitary system (Alternative 1) are located inside each floor the acoustical effect of this alternative is slightly more undesirable than acoustical comfort provided with the Alternative 2 which requires all the units to be located on the roof and out of the occupied floors. As a sample the relative weights for water efficiency, energy efficiency, acoustical comfort, and pollution have been shown below in Table 9.13. Please remember that in AHR method instead of the traditional method of calculating eigenvector we have used another one of the possible methods that Saaty has recommended. As one of the acceptable approximations of the eigenvector in AHR method, Saaty proposes that the eigenvector of a n x n matrix can be calculated with multiplication of each row of the matrix, and then taking the nth root of the result of the multiplication. These numbers then need to be normalized to sum up to 1. The normalized single column will be a very good approximation of the eigenvector of the matrix. Assume that the first row of a 4×4 AHR evaluation matrix is consisted of (1, 1/3, 1/5, 1/9). In AHR method $(1 \times 1/3 \times 1/5 \times 1/9)^{0.25} = 0.293$. This value and the other three values generated from similar processes for second, third, and fourth row of the original matrix create a single column (4×1) matrix which after normalization will represent the eigenvector of the 4×4 matrix. Another frequently used method is utilizing a higher power of the matrix (e.g., power of 2) to come up with the priority matrix. As the result of this exercise we can see that from the point of view of, e.g., acoustical comfort Alternative 2 has advantage over Alternative 1.

Table 9.13 Water and energy efficiency, acoustical comfort, and pollution matrices for Alt1 and Alt 2

Water efficiency				
	Alt 1	Alt 2	Water efficiency	normalized
Alt 1	1.00	0.33	0.57	0.25
Alt 2	3.00	1.00	1.73	0.75
			2.31	1.00

Energy efficiency				
	Alt 1	Alt 2	Energy efficiency	Normalized
Alt 1	1.00	5.00	2.24	0.83
Alt 2	0.20	1.00	0.45	0.17
			2.68	1.00

Acoustical comfort				
	Alt 1	Alt 2	Acoustical comfort	Normalized
Alt 1	1.00	0.33	0.58	0.25
Alt 2	3.00	1.00	1.73	0.75
			2.31	1.00

Pollution				
	Alt 1	Alt 2	Pollution	Normalized
Alt 1	1.00	0.33	0.58	0.25
Alt 2	3.00	1.00	1.73	0.75
			2.31	1.00

Performing energy modeling with similar parameters except HVAC systems that are used for Alternatives 1 and 2 provides an insight into respective energy consumption, quantity of pollution, and water consumption of the alternatives. We can then use these calculated quantities to come up with their relative priority as it can be seen below.

At this point we have gathered all the information that we need in order to complete our analytical hierarchy process. With comparison of the sum of the multiplication of all the parameters' priority by the main matrix that we developed earlier gives us two normalized values that shows which alternative has advantage over the other.

Similar method can be used for comparison of three or more alternatives. Also more detail subcategories can be assigned to the site, renewable energy, thermal comfort, and daylight and view parameters (Table 9.14).

Table 9.14 Overall evaluation of Alt 1 vs. Alt 2

	Site	Water efficiency	Energy and atmosphere		Material use	Indoor air quality					Result
			Energy efficiency	Renewable energy		Outside air	Acoustical comfort	Thermal comfort	Day light and view	Pollution	
	0.22	0.07	0.24	0.24	0.06	0.37	0.37	0.37	0.37	0.37	
	1	1	0.88	0.12	1	0.33	0.04	0.14	0.14	0.33	
Alt 1	0.50	0.25	0.83	0.50	0.50	0.50	0.25	0.50	0.50	0.25	
Alt 2	0.50	0.75	0.17	0.50	0.50	0.50	0.75	0.50	0.50	0.75	
Water cooled unitary	0.11	0.0175	0.175296	0.0144	0.03	0.06105	0.0037	0.0259	0.0259	0.030525	0.494271
Air cooled unitary	0.11	0.0525	0.035904	0.0144	0.03	0.06105	0.0111	0.0259	0.0259	0.091575	0.458329

References

1. Saaty, T. L. (2008). Decision making with the analytic hierarchy process. *International Journal of Services Sciences, 1*(1).
2. Triantaphyllou, E., & Mann, S. H. (1995). Using the analytic hierarchy process for decision making in engineering applications: Some challenges. *International Journal of Industrial Engineering: Applications and Practice, 2,* 35–44.
3. Sloane, E. B., Liberatore, M. J., Nydick, R. L., Luo, W., & Chung, Q. B. (2003). Using the analytical hierarchy process as a clinical engineering tool to facilitate an iterative, multidisciplinary, microeconomic health technology assessment. *Computers and Operations Research, 30*(10), 1447–1465.
4. Kousalya, P., Reddy, G. M., Supraja, S., & Prasad, V. S. (2012). Analytical hierarchy process approach – an application of engineering education. *Mathematica Aeterna, 2*(10), 861–878.
5. Coyle, G. *The analytical hierarchy process; practical strategy. Open access material.* AHP. http://www.booksites.net/download/coyle/student_files/AHP_Technique.pdf

Chapter 10
Genetic Algorithm Optimization

Abstract In a daily basis the HVAC and architectural engineering professionals are faced with conditions that they need to make complex decisions while satisfying multiple objectives that may also be conflicting as well: decisions such as how to improve the design building element selection to minimize the cost and maximize the comfort level, how to organize the tasks in hands in optimum order to maximize the effectiveness of the operation while minimizing the consumed time, how to improve the firm design check lists and orderly follow these design check-list tasks to make sure that no major task is left unfulfilled while not hurting the project delivery schedule, and how to reach an agreement/compromise with other team members to make sure that all trades' targets such as electrical engineer desire to maximize possible daylighting, architect needs to maximize the size of glazing surfaces, and the mechanical engineer yearning to consume less energy are satisfied.

Keywords Genetic algorithm optimization • Chromosome • Gene • Binary values • Weighted sum approach • Altering objective functions • Pareto-ranking approach • Tournament selection • Rank-based roulette wheel selection • Steady-state selection • Proportional Roulette wheel selection • Mutation

In a daily basis the HVAC and architectural engineering professionals are faced with conditions that they need to make complex decisions while satisfying multiple objectives that may also be conflicting as well: decisions such as how to improve the design building element selection to minimize the cost and maximize the comfort level, how to organize the tasks in hands in optimum order to maximize the effectiveness of the operation while minimizing the consumed time, how to improve the firm design check lists and orderly follow these design check-list tasks to make sure that no major task is left unfulfilled while not hurting the project delivery schedule, and how to reach an agreement/compromise with other team members to make sure that all trades' targets such as electrical engineer desire to maximize possible daylighting, architect needs to maximize the size of glazing surfaces, and the mechanical engineer yearning to consume less energy are satisfied.

Another one of the methods that has been used under general umbrella of both single- and multi-criteria decision-making problem solving and can drastically improve such tasks is genetic algorithm optimization. As it can be guessed from its name, genetic algorithm optimization has gained its designation and also its process pattern from the genetic science in biology and natural science. In nature two chromosomes that are built of multiple genes interact with each other and exchange some genes. This process when is repeated multiple times can cause creation of newer and more advanced chromosomes that leads to evolutionary improvement of the original specie. Similarly in a genetic algorithm optimization process the decision maker starts by randomly choosing some building elements for systems that he or she is pursuing to optimize. He or she then changes these building elements systematically and each time checks the strength of the system against one (or multiple and generally conflicting) objective(s) and improves the system as a whole, further and further until he or she reaches an optimal solution for his or her problem question. In other words, in this approach the decision maker will provide some systems (chromosomes) made of building blocks (genes) and tries to cross over the systems (chromosomes) on a randomly selected pairwise pattern with each other in order to improve the original system and advance its components (genes). For example assume system A is assigned a set of binary values such as (1, 0, 0, 1, 1, 0, 1, 1) in which parameter numbers 1, 4, 5, 7, and 8 are assigned desirable and high capacity of something good (e.g., and in our field of interest high efficiency, low energy consumption or high level of comfort) while parameter numbers 2, 3, and 6 are assigned undesirable and low capacity of something good (e.g., low durability, high cost). A pool of N systems (chromosomes) then is randomly selected with its associated building block (genes) binary values also randomly assigned. Sum of the values of these building blocks for each system in this case represents its overall strength. For example the strength of the system A described above is equal to 5 and strength of another possible choice system B with binary values of (1, 0, 0, 1, 0, 0, 1, 1) will be equal to 4. As it was pointed out earlier these values are representative of the strength or fitness of each system. In natural science the higher the fitness of the chromosome (system) is, the higher is its chance of survival and passing its characteristics (genes) to its off-springs. In genetic algorithm optimization the proposed process is very similar to that. The basic objective is to improve one or a group of systems to maximum possible or even if it is viable to the complete fitness (all ones and no zeros) condition which will produce the optimized solution. Konak et al. [1] depicted three common methods for setting the fitness function. These methods are weighted sum approach, altering objective functions, and Pareto-ranking approach. Scientists generally use such approaches in order to rank the systems against each other when use this technique.

After selecting our N systems which each one is described by its binary values and therefore relative fitness compared to other systems, the process continues with randomly selecting a number of two system sets which are required to interact one by one to create better possible systems. Different literatures [2, 3] offer different methods for selecting sets of two systems such as tournament and rank-based roulette wheel selection, and steady-state selection, but one of the most popular methods is the proportional roulette wheel selection in which the surface of the roulette will be divided into N slices that each will be covering an area of the roulette equal to the relative fitness of the specific system compared to other systems. For example assume that we have four systems (A (1, 0, 0, 1, 1, 0, 1, 1), B (1, 0, 0, 1, 0, 0, 1, 1), C (1, 1, 1, 1, 0, 0, 1, 1), and D (0, 0, 0, 1, 0, 0, 0, 0)) and the fitness of each section is 5, 4, 6, and 1, respectively. This makes a total roulette area of 16 units, and therefore the assigned area of each system on the roulette surface will be equal to 5/16, 4/16, 6/16, and 1/16, respectively. This will cause a higher chance of being selected and therefore becoming paired against other systems for system C. The chances of the other three systems to be selected will be smaller than system C and in descending order will be system A, B, and D.

After assigning the partial chances to each system a random draw based on the relative fitness of each system selects and matches multiple two system sets in order to develop new and improved system sets and replace the original system sets in the space with the new ones. For example in our four system space assume that the first two system set is C and B and the second two system set is C and A selected completely on a random basis just due to their higher probability versus system D. When C and A are chosen to interact or as it is called in this method "to cross over," one option is that the off-springs of this crossover action will get value of 1 where both parents are assigned fitness 1, and will get 0 where both parents are assigned 0, but where one of the parents is assigned 1 and the other one is assigned 0 fitness, the off-springs will receive either 1 or 0 on a random basis. It should be noted here that this is just one of the methods for assigning value to the off-springs. Therefore the first off-spring of C and A, or E, will probably shape up as E (1, 1, 0, 1, 1, 0, 1, 1) and the second off-spring or F will probably shape up to be F (1, 1, 0, 1, 0, 0, 1, 1) and as C and B interact or cross over, the off-springs will be G (1, 0, 1, 1, 0, 0, 1, 1) and H (1, 1, 0, 1, 0, 0, 1, 1). Based on this procedure the new four system space will be E, F, G, and H, instead of the original A, B, C, and D, and the total system fitness will be changed to sum of 6, 5, 5, and 5 or 21 which in comparison to the original system fitness of 5, 4, 6, and 1 or 16 is an obvious advancement. It means that we have used the strengths of the original system elements to improve the system elements' weaknesses and as a result we have created a better overall space. Now the fitness of elements of the system are 6/21, 5/21, 5/21, and 5/21 and another round of algorithm will be performed to reach to a system closer to all ones or the best possible solution. Another more commonly used method is called simple one-point crossover function. In this method each two chromosomes splice at one specific point and then exchange values above and below that point to create two new chromosomes replacing the original two in the chromosome space. Another crossover types include two-point, uniform, and random crossover. Randomly in order to improve the process of solution finding and to locate the best solutions we purposely change one of ones to zero or one of zeros to one in gene level. This action in genetic algorithm is called mutation.

From the point of view of this methods' applicability we can mention where Wall [4] utilized a genetic algorithm for finding optimal solutions to dynamic resource-constrained scheduling problems in manufacturing sites. Such method can be easily implemented for any other resource-constrained scheduling problem including design process or construction process in the building industry as well. In his words "In its most general form, the resource-constrained scheduling problem asks the following: Given a set of activities, a set of resources, and a measurement of performance, what is the best way to assign the resources to the activities such that the performance is maximized?" [4]. Such problems perfectly fit to the realm of genetic algorithm approach for optimization. Akachukwu et al. [5] showed "GA (genetic algorithm) is a type of evolutionary algorithm (EA) that is found useful in so many engineering applications which includes numerical and combinatorial optimization problems, filter design as in the field of Signal processing, designing of communication networks, semiconductor layout, spacecraft [6, 7] and so on."

Now that we are somehow familiar with the structure of genetic algorithm let's try to adopt this method in order to generate a solution for a typical HVAC system selection and discuss this in a more detail with proper steps that should be taken to perform an optimization process with this algorithm. To do so let's refer back to our multi-story building that we have been using as an example throughout this book. Let's assume through an analytical hierarchy process as it was explained in earlier chapter, we have come to the conclusion that the best type of HVAC system for such building is a water source heat pump cooling and electric resistance strip heating systems. Let's also assume we want to use a single-objective genetic algorithm approach to calculate the best overall solution (single objective) from the point of view of the building energy consumption and occupants' comfort level. Even though it looks like two objectives we can change it to a single-objective optimization problem by dividing the energy consumption by capacity of outside air (as the representative of occupant comfort (it can be assumed that the higher the quantity of outside air is in a building the higher the level of comfort of the occupant will be)) to create a single fitness factor. Also assume the main parameters that we want to include in our modeling calculations are quantity of outside air, percent of area of the total glazing system to the total area of the exterior walls, glazing shading coefficient, glazing U-value, wall U-value, roof U-value, and average ceiling height. Other than these factors the rest of the input to the energy modeling software will remain the same for all different conditions. After these primary assumptions let's start the optimization process with considering four different options and randomly populate the above selected parameters. Also assume due to market structure, availability, and price of each element we have hypothetically defined an upper and a lower margin for each of the parameters such that quantity of outside air can be selected between 4 and 10 cfm per person (in each occasion we will add another 0.06 cfm per square foot of the building to calculate the total outdoor air in each case), percent of area of the total glazing system to the total area of the exterior walls can be selected between 40 and 70%, glazing shading coefficient can be selected between 0.16 and 0.30, glazing U-value can be selected between 0.389 and 0.72 (Btu/h ft^2 °F), wall U-value can be selected between 0.039 and 0.091 (Btu/h ft^2 °F), roof U-value can be selected between 0.0385 and 0.07 (Btu/h ft^2 °F), and average ceiling height can be selected between 8.9 and 11 ft.

Let's here note another method that makes our adopted approach much closer to the traditional process of genetic algorithm recommended in literatures. In order to make a better distribution of input parameter instead of selecting the real values or even only 1 or 0 binary option, we can use a four-digit binary parameter for each factor (gene). A four-digit combination of zero's and one's creates an opportunity to develop $(2^4 - 1 = 15)$ intervals that can be used to divide each parameter (gene) span to 15 interval as well. For example if parameter X can be selected in a span between 0.15 and 0.3, instead of assigning 0 to any quantity of X smaller than 0.23 and 1 to any quantity of X above 0.23 or any other selected logic, now we can assign a binary value to each of the possible 15 interval between 0.15 and 0.3. By doing this a combination of 1,0,0,1 (equal to 9) can be the representative of $((9 \times (0.3 - 0.15)/15) + 0.15)$ an X value equal to 0.24, and similarly a combination of 0,0,1,1 (equal to 3) can be the representative of $((3 \times (0.3 - 0.15)/15) + 0.15)$ an X value equal to 0.18. This gives an opportunity for a better distribution of the input values and therefore a higher accuracy in our approach. We also can increase the accuracy if needed by changing each input parameter to an eight-digit binary gene which can create $(2^8 - 1 = 255)$ intervals to work with. Applying this approach to our problem can help us to depict the values in Table 10.1 to a binary combination as it is shown below, and therefore show our chromosomes 1 through 4 as it is shown in Table 10.2 below.

Since my intent here is just to show the process and not to follow a detailed calculation and to the benefit of saving time I would not continue to switch between binary and real values back and forth, and continue with working only with real values because seeing the real numbers may have a higher potential of making it more comprehendible for the reader (Table 10.3).

Using inputs from Table 10.1 below and also various selected outcomes from energy modeling performed for each option through one of the available energy modeling programs (e.g., Trane Trace 700) will lead to the following format where we call option 1, 2, 3, and 4 chromosome 1, 2, 3, and 4, respectively, where the vertical column elements represent each chromosome's gene configuration. See Table 10.4 below, where first seven vertical items in each option are what have been fed to the energy modeling software and the last five vertical items are what energy modeling software generated for each case.

At this point let's try to continue solving the problem with single-objective approaches. To pursue a single-objective approach as it was said earlier let's target the system overall fitness as the target of the optimization which is based on the relation between the building energy consumption and the occupant comfort.

Table 10.1 Selected chromosomes (options)

	Option 1	Option 2	Option 3	Option 4
Outside air quantity (cfm/person)	5	6	7	8
Glazing-to-wall percent	66	60	55	50
Glazing SC	0.29	0.24	0.19	0.18
Glazing U-value	0.65	0.429	0.4	0.488
Wall U-value	0.084	0.063	0.051	0.042
Roof U-value	0.047	0.045	0.052	0.068
Ceiling height (feet)	9	9.5	10	10.5

Table 10.2 Selected chromosome (options) binary and real values

		MIN	MAX	Multiplier		
Outside air		3.5	11	0.5		
Glazing/wall ratio		0.4	0.7	0.02		
SC		0.16	0.3	0.009333		
Glazing U-Value		0.389	0.72	0.022067		
Wall U-Value		0.039	0.091	0.003467		
Roof U-Value		0.0385	0.07	0.0021		
Ceiling height		8.9	11	0.14		
Chromosome 1				Chromosome 2		
Binary	Binary			Binary	Binary	
Input	Input	Real value		Input	Input	Real value
0,0,1,1	3	5		0,1,0,1	5	6
1,1,0,1	13	0.66		1,0,1,0	10	0.6
1,1,1,0	14	0.29		1,0,0,1	9	0.24
1,0,1,0	12	0.65		0,0,1,0	2	0.429
1,1,0,1	13	0.084		0,1,1,1	7	0.063
0,1,0,0	4	0.047		0,0,1,1	3	0.045
0,0,0,1	1	9		0,1,0,0	4	9.5
Chromosome 3				Chromosome 4		
Binary	Binary			Binary	Binary	
Input	Input	Real value		Input	Input	Real value
0,1,1,1	7	7		1,0,0,1	9	8
0,1,1,1	7	0.55		0,1,0,1	5	0.5
0,1,0,0	4	0.19		0,0,1,1	3	0.18
0,0,0,1	1	0.4		0,1,0,0	4	0.488
0,1,0,0	4	0.051		0,0,0,1	1	0.042
0,1,1,0	6	0.052		1,1,1,0	14	0.068
1,0,0,0	8	10		1,0,1,1	11	10.5

Table 10.3 Chromosome (options) binary presentation

Chromosome 1							
0,0,1,1	1,1,0,1	1,1,1,0	1,0,1,0	1,1,0,1	0,1,0,0	1,1,0,1	0,0,0,1
Chromosome 2							
0,1,0,1	1,0,1,0	0,0,1,0	0,0,1,0	0,1,1,1	0,0,1,1	1,0,1,0	0,1,0,0
Chromosome 3							
0,1,1,1	0,1,1,1	0,1,0,0	0,0,0,1	0,1,0,0	0,1,1,0	0,1,1,1	1,0,0,0
Chromosome 4							
1,0,0,1	0,1,0,1	0,0,1,1	0,1,0,0	0,0,0,1	1,1,1,0	0,1,0,1	1,0,1,1

Table 10.4 Building block (gene) values for the selected systems (chromosomes)

	Chromosome 1	Chromosome 2	Chromosome 3	Chromosome 4
Outside air (cfm/p)	5	6	7	8
Glazing-to-wall percent	66	60	55	50
Glazing SC	0.29	0.24	0.19	0.18
Glazing U-value	0.65	0.429	0.4	0.488
Wall U-value	0.084	0.063	0.051	0.042
Roof U-value	0.047	0.045	0.052	0.068
Ceiling height (ft)	9	9.5	10	10.5
Building energy consumption (Btu/ft^2-year)	24,689	17,935	15,959	17,538
CO_2 generation (lbm/year)	1,274,000	925,000	823,000	905,000
Water consumption (1000 gal)	820	702	642	636
Source energy consumption (Btu/ft^2-year)	74,073	53,811	47,882	52,620
Electricity consumption (kWh)	908,505	659,991	587,273	645,378

Table 10.5 Fitness and ranking of the selected systems (chromosomes)

	Option 1	Option 2	Option 3	Option 4
Outside air quantity/building energy consumption	0.000202519	0.000334541	0.000438624	0.000456152
Fitness	0.141440213	0.23364488	0.306336512	0.318578395
Rank	4	3	2	1

Therefore in order to be able to compare our four system options we define the fitness factor which is constructed of dividing total quantity of outside air per person (as selected representative of occupant comfort) by total building energy consumption. As it was stated before, it should be obvious to the reader that the higher the quantity of outside air provided for occupants the higher the air quality and therefore the higher will be the occupant comfort level.

Using the information from Table 10.4 will lead us to the findings in Table 10.5 above that represents the rank of our four systems based on our defined fitness factor. Therefore the two best options will be option 4 and option 3. There are different approaches available for us in order to continue the process of genetic algorithm to the next step. One of the common methods is to select the best one or two options (in order to keep the available best options in the mix) and then cross over the options based on their fitness level. This is called Elitist method. In order to continue our approach based on this method, we can keep options 4 and 3 as it is, and then randomly select two crossovers (e.g., crossover option four with option 3 and option 4 with option 2) to complete the next set of our genetic algorithm chromosomes.

In processing the crossover function between option 4 and option 3, and option 4 and option 2, we can select simple one-point crossover method. In this method the off-spring of the crossover adopts the characteristic of one of the parents before and up to the selected point, and from that point on it will adopt the characteristics of the other parent. In order to get a better solution let's mutate the second gene of one of the off-spring chromosomes and change it from 50 % glazing-to-wall ratio to 55 % (see Table 10.6 below). At this point with these new information we should perform another set of energy modeling (in this case only two energy modeling is required since the original options 4 and 3 have been left unchanged).

This time as the result of the fitness objective (outside air quantity/building energy consumption) option 8 (original option 4) and option 7 (original option 3) and option 5 are the best solutions. See Table 10.7.

Table 10.6 Crossover of the main chromosomes and the new off-springs

	Crossovers 4 and 3	Crossovers 4 and 2	Keep 3	Keep 4
	Chromosome 5	Chromosome 6	Chromosome 7	Chromosome 8
Outside air (cfm/p)	8	6	7	8
Glazing-to-wall percent	55	60	55	50
Glazing SC	0.18	0.24	0.19	0.18
Glazing U-value	0.488	0.429	0.4	0.488
Wall U-value	0.051	0.042	0.051	0.042
Roof U-value	0.052	0.068	0.052	0.068
Ceiling height (ft)	10	10.5	10	10.5
Building energy consumption (Btu/ft^2-year)	18,283	17,983	15,959	17,538
CO_2 generation (lbm/year)	943,692	928,201	823,000	905,000
Water consumption (1000 gal)	646	702	642	636
Source energy consumption (Btu/ft^2-year)	54,853	53,953	47,882	52,620
Elec consumption (kWh)	672,773	661,729	587,273	645,378

Table 10.7 Fitness of the off-springs

	Option 5	Option 6	Option 7	Option 8
Outside air quantity/energy consumption	0.000437565	0.000333648	0.000438624	0.000456152
Fitness	0.262645648	0.20027041	0.26328132	0.273802622
Rank	2	4	3	1

This process should be repeated until we reach the best possible solution in which our objective of relative value of the building energy consumption and occupant comfort reaches an optimum value.

As it was said earlier we can also optimize the whole chromosome based on all the genes as well. In this method instead of the relative value of the building energy consumption and occupant comfort of the previous method we need to evaluate all the elements together. One possible way is to go back to Table 10.1, compare each gene of every chromosome, and assign zero to those values that are considerable weaker than the same values in the other systems, and assign 1 to the values that are considerably stronger than the rest of the same values in the same category. If we believe not all the genes should have the same weight in our system, we also can generate a relative value (sum of all values should add up to 1) and assign this relative multiplier value to each gene. Multiplying this weight value to each horizontal line and then adding all the numbers in each vertical line would represent the overall fitness of each system.

Now we can rank the best systems and cross over them in a pairwise basis to generate new off-springs of our original systems. The process should be continued until we get to a system with satisfactory fitness value. See Tables 10.8, 10.9, and 10.10.

Table 10.8 Complete chromosome fitness method

Normalized				
Option 1	Option 2	Option 3	Option 4	
Water source HP	Water source HP	Water source HP	Water source HP	System cooling type
ELECTRIC	ELECTRIC	ELECTRIC	ELECTRIC	System heating type
0.19	0.23	0.27	0.31	Outside air quantity (cfm)
0.29	0.26	0.24	0.22	Glazing-to-wall percent
0.32	0.27	0.21	0.20	Glazing SC
0.33	0.22	0.20	0.25	Glazing U-value
0.35	0.26	0.21	0.18	Wall U-value
0.22	0.21	0.25	0.32	Roof U-value
0.32	0.24	0.21	0.23	Building energy consumption (Btu/ft^2-year)
0.32	0.24	0.21	0.23	CO$_2$ generation (lbm/year)
0.29	0.25	0.23	0.23	Water consumption (1000 gal)
0.23	0.24	0.26	0.27	Ceiling height (ft)
0.32	0.24	0.21	0.23	Source energy consumption (Btu/ft^2-year)
0.32	0.24	0.21	0.23	Elec consumption (kWh)

Table 10.9 Complete chromosome fitness method: Binary presentation

Option 1	Option 2	Option 3	Option 4	
Water source HP	Water source HP	Water source HP	Water source HP	System cooling type
ELECTRIC	ELECTRIC	ELECTRIC	ELECTRIC	System heating type
0	0	1	1	Outside air quantity (cfm)
1	1	0	0	Glazing-to-wall percent
1	1	0	0	Glazing SC
1	0	0	1	Glazing U-value
1	1	0	0	Wall U-value
0	0	1	1	Roof U-value
0	1	1	1	Building energy consumption (Btu/ft^2-year)
0	1	1	1	CO_2 generation (lbm/year)
0	0	1	1	Water consumption (1000 gal)
0	0	1	1	Ceiling height (ft)
0	1	1	1	Source energy consumption (Btu/ft^2-year)
0	1	1	1	Elec consumption (kWh)

Whitley [8] recognizes two interpretations of genetic algorithm: one as the canonical model that was developed by John Holland and his students in mid-1970s and is the common procedure for genetic algorithm as it is described here, and the other one as any population-based model that utilizes selection and recombination operators to generate new sample points in a search space. He also concludes that "One thing that is striking about genetic algorithms and the various parallel models is the richness of this form of computation. What may seem like simple changes in the algorithm often result in surprising kinds of emergent behavior. Recent theoretical advances have also improved our understanding of genetic algorithms and have opened the door to using more advanced analytical methods."

For an extensive list of applications of genetic algorithm in different fields of research see Wikipedia page https://en.wikipedia.org/wiki/List_of_genetic_algo rithm_applications.

Table 10.10 Complete chromosome fitness method and the ranking

		Option 1	Option 2	Option 3	Option 4
		Water source HP	Water source HP	Water source HP	Water source HP
	Relative value	ELECTRIC	ELECTRIC	ELECTRIC	ELECTRIC
Higher better (more comfortable)	0.2	0	0	0.2	0.2
Higher better (more beautiful)	0.15	0.15	0.15	0	0
Higher better (cheaper)	0.1	0.1	0.1	0	0
Higher better (cheaper)	0.05	0.05	0	0	0.05
Higher better (cheaper)	0.025	0.025	0.025	0	0
Higher better (cheaper)	0.025	0	0	0.025	0.025
Lower better	0.1	0	0.1	0.1	0.1
Lower better	0.05	0	0.05	0.05	0.05
Lower better	0.05	0	0	0.05	0.05
Higher better (more beautiful)	0.05	0	0	0.05	0.05
Lower better	0.1	0	0.1	0.1	0.1
Lower better	0.1	0	0.1	0.1	0.1
	1				
		0.325	0.625	0.675	0.725
NORMALIZED		0.138297872	0.265957447	0.287234043	0.308510638
Rank		4	3	2	1

References

1. Konak, A., Coit, D. W., & Smith, A. E. (2006). Multi-objective optimization using genetic algorithms: a tutorial. *Reliability Engineering and System Safety, 91*(9), 992–1007.
2. Razali, N. M., & Geraghty, J. (2011). Genetic algorithm performance with different selection strategies in solving TSP. *Proceedings of the World Congress on Engineering, 1*, 156.
3. Malhotra, R., Singh, N., & Singh, Y. (2011). Genetic algorithms: Concepts, design for optimization of process controllers. *Computer and Information Science, 4*(2), 39.
4. Matthew Bartschi Wall. (1996). A genetic algorithm for resource-constrained scheduling.
5. Akachukwu, C. M., Aibinu, A. M., Nwohu, M. N., & Salau, H. B. (2014). A decade survey of engineering applications of genetic algorithm in power system optimization. *5th International Conference on Intelligent Systems, Modelling and Simulation.*
6. OlesyaPeshko. (2007). *Global optimization genetic algorithms.* McMaster University Hamilton, Ontario ppt presentation 2007 (p. 25).
7. Bhattacharjya, R. K. (2012, October 19). *Introduction to genetic algorithms.* Department of Civil Engineering. Indian Institute of Technology Guwahati.
8. Whitley, D. *A genetic algorithm tutorial.* http://ir.nmu.org.ua/bitstream/handle/123456789/115124/fb354ca780d35ffcf82cc1b44a5a6c35.pdf?sequence=1

Chapter 11
Pareto-Based Optimization

Abstract Engineering and architectural problems most of the times are not dealing with only one single objective to be satisfied. To the contrary almost always these problems contain different objectives to be fulfilled: problems such as how to select the optimized option among objectives such as energy consumption, comfort, and beauty of the building, or construction cost, operation and maintenance costs, and expected life-span of the design building.

Keywords Pareto optimal • Non-dominated • Dominated • A priori • A posteriori • Progressive • Evolutionary • Thermal comfort • Comfort indicator • Pareto front • Pareto frontier

Pareto Domination

Engineering and architectural problems most of the times are not dealing with only one single objective to be satisfied. To the contrary almost always these problems contain different objectives to be fulfilled: problems such as how to select the optimized option among objectives such as energy consumption, comfort, and beauty of the building, or construction cost, operation and maintenance costs, and expected life-span of the design building. For solving multi-objective optimization problems it is common to find a way to combine multiple objectives into one single objective and solve the optimization problem like a single-objective problem (similar to what was explained in genetic algorithm in the earlier section), move all the objectives except one to the constraint set, or use the Pareto optimal front of multiple solutions.

The first two methods can provide a single best answer for the problem, but the third method is capable of providing a set of good (non-dominated) solutions that decision maker then can choose the best fit for his or her application among them. In real-life engineering problems the application of Pareto optimal method is usually more practical than other methods because most of the time we need to make some trade-off between multiple decision makers who are involved in the problem-solving activity. This method is named after the Italian economist/mathematician Vilfredo Pareto (1848–1923) who originally invented the process. The process of finding Pareto optimal solution is in fact trying to find the non-dominated solutions

© Springer International Publishing Switzerland 2016

J. Khazaii, *Advanced Decision Making for HVAC Engineers*,

DOI 10.1007/978-3-319-33328-1_11

among the possible solutions for the optimization problem. If a solution serves multiple objectives, a non-dominated solution is a solution that we cannot make improvement on any of the objectives without lowering the value of at least one of the other solutions' objectives. With two objective optimization which is the simplest level of multi-objective optimization we can create a depiction of Pareto frontier (Pareto front) in the x–y space coordination with each axis representing one of the optimization objectives. When we are dealing with the problems with more than two objectives in real life it becomes harder and impractical to illustrate the solution similar to the way in which it can be presented for two objective Pareto optimization problems. Deb and Kanpur [1] in their paper have presented an evolutionary method for problems with multiple objectives (about ten objectives). Meanwhile they have stated basic reasons that solving problems with more than two or maximum three objectives may not be effective from the practical perspective. The reasons are the following: "First, visualization of a large-dimensional front is certainly difficult. Second, an exponentially large number of solutions would be necessary to represent a large-dimensional front, thereby making the solution procedures computationally expensive. Third, it would certainly be tedious for the decision-makers to analyze so many solutions, to finally be able to choose a particular region, for picking up solution(s)" [1].

In general multi-objective optimization can be categorized to non-Pareto-based and Pareto-based optimizations. A good example of non-Pareto-based multi-objective optimization method is based on game theory which will be discussed in more detail in a later chapter of this book. Even though different categorizations have been noted for Pareto-based optimization, one most recognized method is to subcategorize Pareto optimization to "a priori" and "a posteriori" Pareto, "progressive" and "evolutionary." [3] In a priori method adequate primary information should be developed before the optimization process, but in a posteriori method it is required that a large number of Pareto optimal solutions to be generated first and then the designer should choose his or her best trade-off case among them. A progressive method is a combination of these two earlier mentioned options [4]. Evolutionary methods are named after the general biology evolution theory and are based on the survival of the fittest solutions. One of the most commonly used evolutionary methods is multi-objective genetic algorithm in which in each iteration a group of new solutions are generated. By using this method we can develop many possible answers at each iteration based on multiple factors and then develop the Pareto front and select the solution that fits the best from that group of the solutions. In a multi-objective genetic algorithm optimization method the fit and non-dominant solutions generate a Pareto front, which will take place of the weakest solutions of the first iteration and iteration continues until the best solutions are appeared. The main advantage of using a Pareto optimization therefore is that it filters the researcher's choice down to a smaller group of options that he or she needs to evaluate in more depth while trying to make the final decision. This group of options (solutions) are not dominated by any other option and therefore are the best set of solutions for the decision maker to choose among them.

As an example let's go back to the same building that we have been using in earlier chapters of this book, and assume that a specific system has been selected to cool and heat the building. Also assume that we have selected some random input for some of the building design parameters, have run eight different energy modeling, and have gathered the selected output for each system as it is shown in Table 11.1 and the normalized values of the same output for each system as it is shown in Table 11.2. Now the way that a Pareto optimization works is to compare every two pairs of options and find out if there is any solution that is not dominated by any other solution, and removes all the other dominated solutions from the pool of possible selection choices. The way that a non-dominated solution is defined is that it should be at least equal to the other solution in all criteria except that at least in one criteria it should be better than the other option. Therefore we need to perform a lengthy process of pairwise comparison between every two relevant options' outcomes. The pairwise comparison compares the values of the output of each option and score 1 for the better output and zero for the worse option. Tables 11.3, 11.4, 11.5, and 11.6 represent a pairwise comparison between options 1 and 2, 1 and 3, 1 and 4, 1 and 5, 1 and 6, 1 and 7, 1 and 8, 2 and 3, 2 and 4, etc. Such comparison shall be done for all the options. If an option in a pairwise comparison against another option scores a sum value of zero it means it is dominated by the other option. Such option does not belong to the Pareto frontier of the solutions.

If we complete the pairwise comparison and include the results in a table similar to Table 11.7, we can see that when we are investigating all the objectives, then among the eight options that we have selected options 4 and 8 are dominated by other options (option 4 is dominated by options 1 and 5 and option 8 is dominated by options 1, 2, 4, and 7), and the rest of the options are non-dominated solutions. Therefore the Pareto non-dominated set for this optimization will be generated from the other six options. This simplifies the decision maker's decision in selecting a trade-off solution among these six left options. As it was said earlier, representing the Pareto non-dominated set for conditions with more than two objectives is not practical in real-world problems; therefore to provide a presentation of Pareto frontier, let's limit our objectives to only energy consumption and glazing-to-wall percentage. By going back to Tables 11.1 and 11.2 we can find the real and normalized values of these two objectives for all of the eight options.

If we assume a relative importance for energy consumption versus glass-to-wall percent of 0.65–0.35 we can depict the solutions and a Pareto frontier such as the one that has been shown in Fig. 11.1. If we perform the same analysis but instead of one relative importance use three relative importance of 0.6–0.4, 0.65–0.35, and 0.7–0.3 we will have a set of solutions and a Pareto non-dominated set such as the one shown in Fig. 11.2.

Table 11.1 Selected output for options 1 through 8

		Option 1	Option 2	Option 3	Option 4	Option 5	Option 6	Option 7	Option 8	Maximum
Less is better	Building energy consumption (Btu/ft²-year)	12,700	13,500	14,600	12,700	12,300	15,300	13,200	14,000	15,300
Less is better	Source energy consumption (Btu/ft²-year)	38,300	37,300	43,900	38,200	36,900	44,000	39,000	43,000	44,000
Less is better	CO₂ generation (lbm/year)	659,000	612,000	755,000	657,000	636,000	667,000	630,000	700,000	755,000
Less is better	Construction cost ($/ft²)	27	26	25	28	27	24	25	28	28
More is better	Ceiling height (ft)	9	9	9	9	9	10	10	9	10
More is better	Percent of glass/wall (%)	66	66	66	50	50	40	40	40	66

Table 11.2 Selected output for options 1 through 8: Normalized

		Option 1	Option 2	Option 3	Option 4	Option 5	Option 6	Option 7	Option 8
Less is better	Building energy consumption (Btu/ft^2-year)	0.83	0.88	0.95	0.83	0.80	1.00	0.86	0.92
Less is better	Source energy consumption (Btu/ft^2-year)	0.87	0.85	1.00	0.87	0.84	1.00	0.89	0.98
Less is better	CO_2 generation (lbm/year)	0.87	0.81	1.00	0.87	0.84	0.88	0.83	0.93
Less is better	Construction cost ($/ft^2)	0.96	0.93	0.89	1.00	0.96	0.86	0.89	1.00
More is better	Ceiling height (ft)	0.90	0.90	0.90	0.90	0.90	1.00	1.00	0.90
More is better	Percent of glass/wall (%)	1.00	1.00	1.00	0.76	0.76	0.61	0.61	0.61

Table 11.3 Option 1 vs. other 7 options

	Option 1	Option 2	Option 1	Option 3	Option 1	Option 4	Option 1	Option 5	Option 1	Option 6	Option 1	Option 7	Option 1	Option 8
Less is better — Building energy consumption (Btu/ft²-year)	1.00	0.00	1.00	0.00	0.00	0.00	1.00	1.00	1.00	0.00	1.00	0.00	1.00	0.00
Less is better — Source energy consumption (Btu/ft²-year)	0.00	1.00	1.00	0.00	0.00	0.00	1.00	1.00	1.00	0.00	1.00	0.00	1.00	0.00
Less is better — CO$_2$ generation (lbm/year)	0.00	1.00	1.00	0.00	0.00	0.00	1.00	1.00	0.00	0.00	0.00	1.00	1.00	0.00
Less is better — Construction cost ($/ft²)	0.00	1.00	0.00	1.00	1.00	0.00	0.00	0.00	0.00	1.00	0.00	1.00	0.00	0.00
More is better — Ceiling height (ft)	0.00	0.00	0.00	0.00	0.00	0.00	0.00	0.00	0.00	1.00	0.00	1.00	1.00	0.00
More is better — Percent of glass/wall (%)	0.00	0.00	0.00	0.00	1.00	0.00	1.00	0.00	1.00	0.00	1.00	0.00	1.00	0.00
Sum	1.00	3.00	3.00	1.00	2.00	0.00	4.00	3.00	3.00	2.00	3.00	3.00	5.00	0.00

Table 11.4 Option 2 vs. other 6 options

		Option 2	Option 3	Option 2	Option 4	Option 2	Option 5	Option 2	Option 6	Option 2	Option 7	Option 2	Option 8
Less is better	Building energy consumption (Btu/ft^2-year)	1.00	0.00	0.00	1.00	1.00	1.00	1.00	0.00	1.00	0.00	1.00	0.00
Less is better	Source energy consumption (Btu/ft^2-year)	1.00	0.00	1.00	0.00	1.00	1.00	1.00	0.00	1.00	0.00	1.00	0.00
Less is better	CO$_2$ generation (lbm/year)	1.00	0.00	1.00	0.00	1.00	0.00	1.00	0.00	1.00	0.00	1.00	0.00
Less is better	Construction cost ($/ft^2)	0.00	1.00	1.00	0.00	0.00	0.00	0.00	1.00	1.00	0.00	1.00	0.00
More is better	Ceiling height (ft)	0.00	0.00	0.00	0.00	0.00	0.00	0.00	1.00	0.00	1.00	0.00	0.00
More is better	Percent of glass/wall (%)	0.00	0.00	1.00	1.00	1.00	0.00	1.00	0.00	1.00	0.00	1.00	0.00
	Sum	3.00	1.00	4.00	1.00	3.00	2.00	4.00	2.00	5.00	1.00	5.00	0.00

Table 11.5 Option 3 vs. other 6 options

		Option 3	Option 4	Option 3	Option 5	Option 3	Option 6	Option 3	Option 7	Option 3	Option 8
Less is better	Building energy consumption (Btu/ft²-year)	0.00	1.00	0.00	1.00	1.00	0.00	0.00	1.00	0.00	1.00
Less is better	Source energy consumption (Btu/ft²-year)	1.00	0.00	0.00	1.00	0.00	0.00	0.00	1.00	0.00	1.00
Less is better	CO_2 generation (lbm/year)	1.00	0.00	0.00	1.00	0.00	1.00	0.00	1.00	0.00	1.00
Less is better	Construction cost ($/ft²)	0.00	1.00	1.00	0.00	0.00	1.00	0.00	0.00	1.00	0.00
More is better	Ceiling height (ft)	0.00	0.00	0.00	0.00	0.00	1.00	0.00	1.00	0.00	0.00
More is better	Percent of glass/wall (%)	1.00	0.00	1.00	0.00	1.00	0.00	1.00	0.00	1.00	0.00
	Sum	3.00	2.00	2.00	3.00	2.00	3.00	1.00	4.00	2.00	3.00

Table 11.6 Option 4 vs. other 4 options

		Option 4	Option 5	Option 4	Option 6	Option 4	Option 7	Option 4	Option 8
Less is better	Building energy consumption (Btu/ft²-year)	0.00	1.00	1.00	0.00	1.00	0.00	1.00	0.00
Less is better	Source energy consumption (Btu/ft²-year)	0.00	1.00	1.00	0.00	1.00	0.00	1.00	0.00
Less is better	CO_2 generation (lbm/year)	0.00	1.00	1.00	0.00	0.00	1.00	1.00	0.00
Less is better	Construction cost ($/ft²)	0.00	1.00	0.00	1.00	0.00	1.00	0.00	0.00
More is better	Ceiling height (ft)	0.00	0.00	0.00	1.00	0.00	1.00	0.00	0.00
More is better	Percent of glass/wall (%)	0.00	0.00	1.00	0.00	1.00	0.00	1.00	0.00
	Sum	0.00	4.00	4.00	2.00	3.00	3.00	4.00	0.00

Table 11.7 Overall comparison

1	0	0	0	0	0	0	0	0	0
2	0	0	0	0	0	0	0	0	0
3	0	0	0	0	0	0	0	0	0
4	X	0	0	0	X	0	0	0	2
5	0	0	0	0	0	0	0	0	0
6	0	0	0	0	0	0	0	0	0
7	0	0	0	0	0	0	0	0	0
8	X	X	0	X	0	0	X	0	4
Options	1	2	3	4	5	6	7	8	

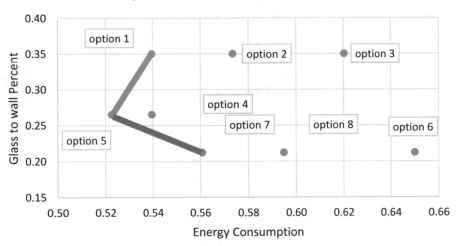

Fig. 11.1 Energy vs. glass-to-wall percent Pareto frontier option 1

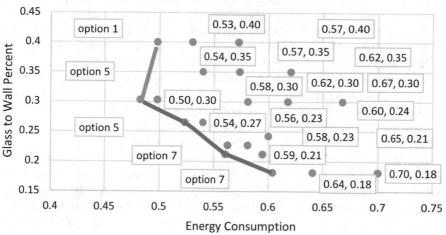

Fig. 11.2 Energy vs. glass-to-wall percent Pareto frontier option 2

Use of Pareto Optimization and Multi-Objective Genetic Algorithm in Energy Modeling

Now that in some degree we have gotten familiar with multi-objective genetic algorithm (in previous chapter) and Pareto domination concepts earlier in this chapter, let's try to put them together and propose a method of optimizing energy consumption of our buildings, beyond what is proposed by ASHRAE 90.1 Standard. Before presenting a method of improvement to the system let's briefly remember what the common ASHRAE 90.1 solution for energy consumption evaluation is. In general ASHRAE 90.1 Standard solution is to perform an energy modeling for the purposed building and compare that with the base building energy modeling result and if the proposed building consumes (costs) less energy than the base building it passes the standard requirement. With ASHRAE 90.1 approach, the process stops here.

Let's assume that we have done our exercise and the ASHRAE 90.1 Standard base building setup for our building leads us to a threshold of 61,000 kbtu/ft^2-year (here as an experiment let's refer to the source energy consumption as indicator instead of building consumed energy cost used by ASHRAE 90.1 Standard which in fact also is built upon calculated building energy consumption). Therefore if our designed building passes the threshold we will satisfy the standard requirement. But why should we stop here. As a responsible engineer dedicated to selecting the best possible choices, let's structure a process to optimize the energy consumption of our building beyond the concept of only passing the threshold.

I think it would make sense to say designing a building only based on the objective of minimizing its energy consumption may not be as exciting since this objective would be easily achievable if we reduce the grade of all the key aesthetical and comfort-related parameters affecting the design of the building. Therefore it would make sense to add another objective to the mix in which we define the values that would contribute not only to the beauty of the building but also to the comfort level of the occupants (of course a little more complicated method than presented in the previous chapter for calculating the comfort-level representative). In doing so first we should select the parameters that we believe might be of some importance in defining an overall comfort factor. We can utilize a similar approach to analytical hierarchy process as we did in earlier chapters in order to develop main categories, and subcategories. Also assume that we define the factors minimum, maximum, and related weights. The first comfort category that we define is the visual comfort factor which will be subcategorized to ceiling height, percent of glazing-to-wall ratio, and level of lighting in the building. Each of these factors has direct effect on the visual comfort of the occupant since higher ceilings, larger percentage of glazing-to-wall ratio, and higher lighting levels all would contribute to the visual comfort of the occupants but in most occasions have diverse effect on the energy consumption of the building. The second category that we would select will be the operational comfort which itself is subcategorized to quantity of outside air and ratio of the available space to number of occupant. In each case people tend to operate better if larger quantity of outside air is delivered to the building and also there is larger space available for each person to live and operate in the building. Finally let's assign the last category and its sole subcategory to thermal comfort and amount of allowed drift in cooling and heating temperature set point. It can be argued that the tighter the drift points from the thermostat set points are for a larger time the temperature would be controlled near the set point and therefore provides higher level of comfort for the occupants (of course the thermal comfort can be calculated with the presented method in ASHRAE 55 Standard, but here for a more simplified method we just rely on temperature sensor allowed drift). Also let's assign the following normalized weights to our main criteria as 0.4, 0.15, and 0.45 for visual comfort, thermal comfort, and operational comfort, respectively. Similarly let's assign the following normalized weights of 0.2, 0.4, and 0.4 to the subcategories of visual comfort (ceiling height, percent of glazing-to-wall ratio, lighting), 1 to subcategory of thermal comfort (drift from temperature sensor set point), and normalized weight of 0.75 and 0.25 to subcategories of operation comfort (outdoor air and ration of available space per person). Putting all these information together we can express the comfort indicator as follows:

$$CI = 0.4 \times (0.2 \times C + 0.4 \times GW + 0.4\,L) - 0.15 \times D + 0.45$$
$$\times\,(0.75 \times OA + 0.25 \times SO) \qquad (11.1)$$

where:

CI = Comfort indicator

C = Ceiling height (Min. 8 ft, Max. 11 ft, or normalized as $(8/11 - 1)$

GW = Glazing-to-wall area ration (Min. 40 %, Max. 80 %, or normalized $(0.5 - 1)$

Table 11.8 System (chromosome) and its building blocks (genes)

Outside air	5
Drift	6
Ceiling height	9
Roof *U*-value	0.047
Wall *U*-value	0.084
Glazing *U*-value	0.65
Glazing shading coefficient	0.29
Space per person	200
Lighting	0.82
Percent of glazing to wall	40
Source energy consumption	58,273
Comfort indicator	0.395

L = Lighting (Min. 0.7 W/ft^2, Max. 1.2 W/ft^2, or normalized as $(0.7/1.2 - 1)$)
D = Drift (Min. 1 °F, Max. 7 °F, or normalized as $(1/7 - 1)$)
OA = Outside air (Min. 4 cfm/person, Max. 10 cfm/person, or normalized as $(0.4 - 1)$)
SO = Space per occupant (Min. 150 ft^2/person, Max. 250 ft^2/person, or normalized as $(150/250 - 1)$)

With an overall minimum and maximum value of 0.29–0.83.

This indicator (objective) is just one that we made based on our judgement for the selected factors and their weight. It should be noted here that other indicators can be provided just as effective and helpful as this one. Now as the other indicator (objective) let's consider the source energy consumption of the building and try to implement our multi (bi)-objective genetic algorithm optimization with the help of Pareto domination approach. Therefore let's reframe our earlier problem as one to find the best solution in a way to maximize the comfort objective that we just developed and at the same time to minimize the building source energy consumption objective.

To set up the problem let's create our main chromosome (system option) based on the system that was proved to be better than ASHRAE 90.1 Standard base line, with a set of associated genes from the same option as follows (Table 11.8):

The first ten genes of this chromosome are used to generate the results for the last two genes through energy modeling and calculation of the comfort indicator. Now let's randomly populate the top ten genes from a pool of values in the ranges as indicated before, and calculate the energy modeling results and also comfort indicator of each chromosome, and develop a total of (e.g.) 12 chromosomes (including the 1 above). Of course the more chromosomes we generate we will have more chance to get to a better result, but at the same time the procedure will become more exhaustive and time consuming. The list of our 12 chromosomes (system options) and their genes and two objectives that are calculated through energy modeling and solving for the comfort (performance) indicator is shown in Table 11.9. Normalized values of the objectives and a Pareto front for the options are shown in Table 11.10 and Fig. 11.3.

Table 11.9 Option 1 through 12

	Option 1	Option 2	Option 3	Option 4	Option 5	Option 6	Option 7	Option 8	Option 9	Option 10	Option 11	Option 12
Outside air	5	6	7	10	5	7	5	10	5	7	7	7
Drift	6	4	5	2	6	4	4	2	4	5	4	2
Ceiling height	9	9.5	10	10.5	9	9	10	10.5	9	9	10	10.5
Roof U-value	0.047	0.045	0.052	0.068	0.047	0.045	0.052	0.068	0.046	0.058	0.052	0.068
Wall U-value	0.084	0.063	0.051	0.042	0.084	0.063	0.051	0.051	0.07	0.084	0.051	0.051
Glazing U-value	0.65	0.429	0.403	0.488	0.65	0.429	0.492	0.488	0.293	0.492	0.383	0.515
Glazing shading coefficient	0.29	0.24	0.19	0.18	0.29	0.24	0.21	0.18	0.33	0.21	0.21	0.327
Space per person	200	180	220	250	200	210	200	250	175	190	220	190
Lighting	0.82	0.9	1	1.05	0.82	0.82	0.85	0.86	0.9	1	1.1	0.96
Percent of glazing to wall	40	60	55	50	66	60	55	46	66	50	55	46
Source energy consumption	58,273	57,333	52,785	60,925	74,073	54,749	54,874	51,882	43,611	60,742	55,059	55,297
Comfort indicator	0.395	0.517	0.556	0.736	0.447	0.549	0.479	0.7	0.447	0.566	0.592	0.586

Table 11.10 Source energy consumption vs. comfort indicator

	Option 1	Option 2	Option 3	Option 4	Option 5	Option 6	Option 7	Option 8	Option 9	Option 10	Option 11	Option 12		Maximum
Source Energy Consumption	74073.00	57333.00	52785.00	60925.00	58273.00	54749.00	54874.00	51882.00	43611.00	60742.00	55059.00	55297.00		74073.00
	1.00	0.77	0.71	0.82	0.79	0.74	0.74	0.70	0.59	0.82	0.74	0.75		Lower Better
Comfort Indicator	0.40	0.52	0.56	0.74	0.45	0.55	0.48	0.70	0.44	0.57	0.59	0.59		0.74
	0.54	0.70	0.76	1.00	0.61	0.75	0.65	0.95	0.60	0.77	0.80	0.80		Higher Better

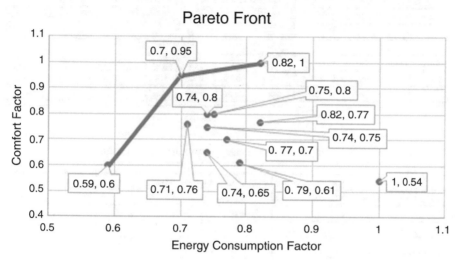

Fig. 11.3 Pareto frontier, energy consumption factor vs. comfort factor

Figure 11.3 shows that systems 8, 9, and 10 are non-dominated systems and create the Pareto non-dominated front. If we are interested we can go to the next best front to add the fourth system (system 3) to the selected systems to be used for the next stage of the optimization.

Now based on multi-objective genetic algorithm let's cross over options 8 and 3 and 9 and 10, by selecting a single point in the length of the chromosomes and generate new chromosomes 8′, 3′, 9′, and 10′ as they are shown in Table 11.11. While doing this crossovers let's do a mutation at the percent of glazing to wall in option 8′. Then perform another set of energy modeling and comfort indicator calculation to develop Table 11.11. As it was discussed earlier mutation process is where in natural genetic evolution and therefore in genetic algorithm as well, a specific gene (building block) changes randomly to a new value which is out of the general order of parental chromosome (structures) procedure.

After normalizing the two objectives in Table 11.11, we will have Table 11.12 which based on its data we can depict the Pareto front in Fig. 11.4. This represents a

Table 11.11 Selected options and its off-springs

	Option 8	Option 9	Option 10	Option 3	Option 8'	Option 9'	Option 10'	Option 3'
Outside Air	10	5	7	7	10	5	7	7
Drift	2	4	5	5	2	4	5	5
Ceiling Height	10.5	9	9	10	10.5	9	9	10
Roof U-Value	0.068	0.046	0.058	0.052	0.068	0.046	0.058	0.052
Wall U-Vlaue	0.051	0.07	0.084	0.051	0.051	0.07	0.084	0.051
Glazing U-Value	0.488	0.293	0.492	0.403	0.403	0.492	0.293	0.488
Glazing Shading Coefficient	0.18	0.33	0.21	0.19	0.19	0.21	0.33	0.18
Space per Person	250	175	190	220	220	190	175	250
Lighting	0.86	0.9	1	1	1	1	0.9	0.86
percent of Glazing to Wall	46	66	50	55	57	50	40	46
Source Energy Consumption	51882	74073	60742	52785	56233	55750	49572	49126
Comfort Indicator	0.7	0.447	0.566	0.556	0.736	0.447	0.549	0.479

Table 11.12 Normalized objectives, source energy consumption vs. comfort indicator

	OPTION8	OPTION9	OPTION10	OPTION3	option 8'	optino 9'	option 10'	option 3'	Maximum
Source Energy Consumption	51882.00	43611.00	60742.00	52785.00	56233.00	55750.00	49572.00	49126.00	60742.00
	0.85	0.72	1.00	0.87	0.93	0.92	0.82	0.81	Lower better
Comfort Indicator	0.70	0.44	0.57	0.56	0.73	0.48	0.48	0.53	0.73
	0.96	0.60	0.78	0.76	1.00	0.66	0.66	0.73	Higher Better

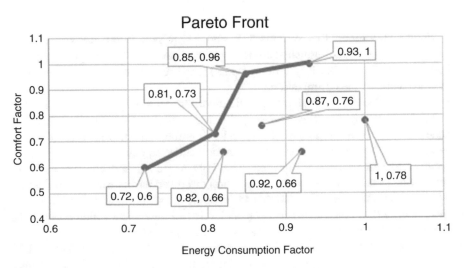

Fig. 11.4 Pareto frontier, energy consumption factor vs. comfort factor

new Pareto front with a set of better solutions from the earlier stage. This time the solutions that create the Pareto non-domination are solutions 8, 9, 3', and 9'.

From the shape and location of the points on the Pareto front solutions 8 and 3 provide the better solutions. Of course the process can continue to generate even better solutions. An interlock between any of the commercially available energy modeling systems with one of the software capable of performing genetic algorithm can drastically help to perform this procedure with more iterations and of course shorter time for reaching the final result, but here my purpose was just to represent a procedure which can improve our way of system selection in our professional approach.

For a list of well-known multi-objective genetic algorithms see Konak et al. [2]. For a list of well-known multi-objective optimization methods see Bakhsh [3].

References

1. Deb, K., & Kanpur, D. K. S. (2005) Through dimensionality reduction for certain large-dimensional multi-objective optimization problems. Genetic Algorithms Laboratory (KanGAL) Indian Institute of Technology Kanpur Kanpur, PIN 208016, India {deb,dksaxena}@iitk.ac.in, http://www.iitk.ac.in/kangal KanGAL Report Number 2005011
2. Konak, A., Coit, D. W., & Smith, A. E. (2006). Multi-objective optimization using genetic algorithms: A tutorial. *Reliability Engineering & System Safety, 91*(9), 992–1007.
3. Bakhsh, A., A., (2013). A posteriori and interactive approaches for Decision-Making with multiple stochastic objectives; PhD thesis; University of Central Florida, Orlando, Florida
4. Hopfe, J.C. (2009). Uncertainty and Sensitivity analysis in building performance simulation for decision support and design optimization; PhD thesis.

Chapter 12
Decision Making Under Uncertainty

Abstract Two different types of uncertainty have been recognized in different literatures. Swiler et al. define epistemic uncertainty as lack of knowledge about the appropriate value to use for quantity (also known as type B, subjective, reducible) which researcher can reduce the level of uncertainty by increasing introduction of the relevant data, and aleatory uncertainty as the type of uncertainty which is characterized by its inherent randomness (type A, stochastic, irreducible) which cannot be reduced by acquiring additional data.

Keywords Uncertainty • Stochastic • Irreducible • Randomness • Reducible • Subjective • Tolerance • Expected utility • Expected value • St. Petersburg dilemma • Expected maximum value • Certainty equivalent • Risk premium • Risk neutral • Risk averse • Risk taker • Bayesian rules • Distribution

Decision Making and Utility Function

Two different types of uncertainty have been recognized in different literatures. Swiler et al. [1] define epistemic uncertainty as lack of knowledge about the appropriate value to use for quantity (also known as type B, subjective, reducible) which researcher can reduce the level of uncertainty by increasing introduction of the relevant data, and aleatory uncertainty as the type of uncertainty which is characterized by its inherent randomness (type A, stochastic, irreducible) which cannot be reduced by acquiring additional data. Kiureghian [2] explains that the word aleatory is derived from Latin word "alea" (rolling dice) and represents the type of uncertainty which is completely random, while epistemic is derived from Greek word "episteme" (knowledge), and represents the type of uncertainty that arises from lack of knowledge. With the help of these definitions, we can express that in building energy modeling we can refer to entities such as weather data or occupant behavior as aleatory uncertainties (those which we generally cannot decrease the level of uncertainty with additional data), while referring to those entities such as thermal characteristics of different material and efficiency of different equipment (that their uncertainty can be decreased by additional collected data) as epistemic uncertainty.

© Springer International Publishing Switzerland 2016
J. Khazaii, *Advanced Decision Making for HVAC Engineers*,
DOI 10.1007/978-3-319-33328-1_12

When we talk about decision making under uncertainty our focus in general is the act of selecting a solution among other possible choices while considering the effects of lack of knowledge (epistemic) about problem inputs. The type of lack of knowledge that we intend to discuss here mostly arises not from modeler's mistakes or underlying mathematics of the solution (model), but those that exist due to the deviation of the selected material or equipment input to the model from those that are actually used in constructing of the real-life object (building). Of course defining and quantifying uncertainties in the input of the model is just one part of the decision making under uncertainty procedure. The other important part is to create a framework to implement the uncertainty that is captured in model output and use it to help a decision maker make the appropriate decision. I will explain this two-step structure in the next few paragraphs in more details in regard to one of the most important functions in building design which is the building energy modeling.

In order to do so, let's first start with defining what are the uncertainties that exist in our energy modeling in a building. For a detail discussion about definition of uncertainty and types of distributions refer to my earlier book "Energy Efficient HVAC design, An Essential Guide for Sustainable Building" [3]. Here I will suffice to a short introduction that I believe is necessary for the reader of this book to follow the rest of the material presented in this chapter.

When in the professional world an energy modeler attempts to define the energy consumption of a building that is going to be built during the design process the general (current) approach is to use one of the commercially available tools such as Trane Trace 700 and eQuest. These tools are perfectly capable of satisfying the required needs depicted by the energy codes and standards such as ASHRAE Standard 90.1 [4] in order to deliver the yearly energy consumption (cost) of the building that is the subject of the design. Under the current established method all the inputs to the energy modeling software are exact quantities, such as U-value of the wall, U-value for the roof and glazing system, shading coefficient of the glazing system, and efficiency of the HVAC equipment. But in fact and due to the tolerances allowed by the testing agencies none of these values are the true values of the items that will be installed in the real building. For example in order to qualify a chiller to have a specific efficiency equal to X kW/t, if the associated testing agency allows $\pm 7.5\%$ deviation when the testing procedure occurs, this allowed tolerance makes all the chillers which their performances fall between $0.925X$ and $1.075X$ to be qualified to be recognized as chillers with X kW/t efficiency. This deviation in values used in model and the actual values associated with the actual material and equipment when is considered as input to the model will result in a probability distribution output of the energy consumption of the building instead of a single value. After this short refresher on the concept of the uncertainty, now let's direct our attention towards the second step of our decision making under uncertainty process, or the framework under which the uncertainty should be assessed to help us make proper decisions.

Until early 1700s the common belief of the scientific community was that people make their decisions to achieve maximum expected value (wealth) from their action. This belief was challenged when in early eighteenth century the St.

Petersburg dilemma (paradox) was presented. St. Petersburg dilemma is a betting game in which the players are invited to participate and bet some amount of money (e.g., 25 gold coins) in return of opportunity to win $2^n (n = 0, 1, 2, \ldots, \infty)$ gold coins. The betting procedure works as follows. The game administer deposits one (2^0) gold coin in a pot and asks the player to throw a fair coin. If in first attempt player gets a tail the game is over and player wins 1 gold coin in place of his or her 25 gold coin bet. If the first throw is head, administer then deposits a total of (2^1) coins in the pot and asks the player to throw the coin again. Similarly if the second throw is a tail the game is over and the player wins 2 gold coins in return of his or her 25 gold coin bet. If the second throw is head again, the game continues, and this time with a total of four (2^2) gold coins in the pot. Therefore if the player continues to make heads without getting a tail on his or her throws his or her winning expectation becomes larger and larger and soon it will be unlimited (e.g., after 10 throws number of the coins in the pot will be 2^{10} (more than 1000 gold coins) and after 20 throws number of the coins in the pot will be 2^{20} (more than 1,000,000 gold coins)). But most of the people in actual life hesitate to participate in this game. This paradox defied the belief that people make decisions to maximize their wealth. The dilemma was partially solved when in mid-1700s the concepts of "value function" and "diminishing marginal value of wealth" were introduced. Even though this solution states that people try to maximize their "value function" and not wealth itself, it was still not capable of describing why people select a sure winning over a gambling choice with the same exact outcome (value function). To understand this question better let's look at two examples. In the first one we assume that we are offered to accept $200 dollars, or instead we have been given the 50–50 chance by throwing a coin to make our winning either $100 or $600. The expected value (function) of staying with our original winning is $200 which is smaller than the expected value (function) of the winning after throwing the coin ($0.5 \times \$100 + 0.5 \times \$600 = \$350$). It is obvious that the reasonable choice is to attempt to win $350 which has more value (function) than $200 winning. But selecting the better choice is not always that obvious. For example assume that we are offered to accept $200 dollars, and we have been given the 50–50 chance by throwing a coin to make our winning either $100 or $300. The expected value (function) of staying with our original offer is $200 which is also the same as the expected value (function) of the winning after throwing the coin ($0.5 \times \$100 +0.5 \times \$300 = \$200$), but almost all the people tend to select the former choice over the latter one. Based on this downfall the value function was not capable of answering the paradox completely either (Charts 12.1 and 12.2).

The question was not answered until in mid-1900s von Neumann and Morgenstern introduced the concept of "utility function" or "attitude of the people toward risk and wealth" and also defined the "expected utility." Based on this model people make their decisions not to get the maximum expected value but to maximize the expected utility which is the combined desire of the decision maker towards wealth and amount of risk that he or she is willing to take. When we are working with value function and utility function we need to understand a few additional main concepts. The first concept is expected utility (EU) which is the

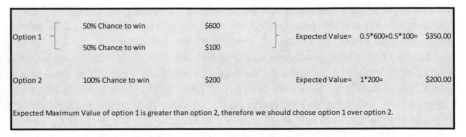

Chart 12.1 Two decisions with different expected values

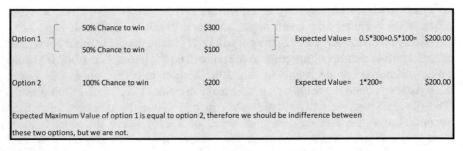

Chart 12.2 Two decisions with similar expected values

weighted average of utilities of all the possible conditions. Expected utility is similar to expected maximum value (EMV) with a difference that the former uses the utility assigned to the values to develop the weighted average, but the latter uses the actual values to do so. The next definition is certainty equivalent (CE) which is the price equivalent of the uncertainty. CE is the price that you need to pay to a risk-averse person to make him or her to take part in an uncertain betting condition, and the price that a risk taker is willing to pay to get into an uncertain betting condition. On a typical risk curve in the first condition CE is to the left of EMV and in the second condition is to the right of EMV. The other notable concept is risk premium (RP) which is the difference between the EMV and CE. For a detail description of utility theory axioms refer to Shoham and Leyton-Brown [5] and Schultz et al. [6].

To get the feeling of how the utility function works, as an example, assume that the satisfaction that you get from the possible option 1 and option 2 of an outcome is 10 and 3, respectively, and the chances that option 1 happens versus option 2 are 40–60. The degree of utility that you get from option 1 will be $(10 \times 0.4 = 4)$ and the degree of utility that you get from option 2 will be $(3 \times 0.6 = 1.8)$, and therefore you will choose option 1 with the higher expected utility.

As another example involving the certainty equivalent concept, let's assume that the risk premium for accepting the option 1 of a decision choice is $60. Option 1 odds are 50–50 to win a maximum amount of $300 and a minimum amount of $100. Also let's assign the utility 1 to the best choice (winning $300) and utility 0 to the worse option (winning $100). The expected utility of option 1 therefore is equal to 0.5. This is also the utility of certainty equivalent (Chart 12.3).

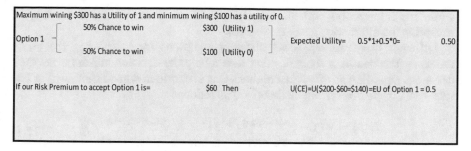

Chart 12.3 Expected utility and risk premium

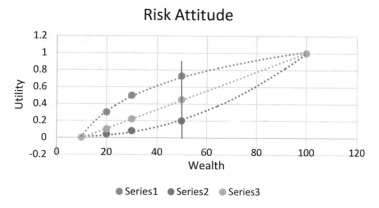

Fig. 12.1 Utility and risk attitude

The utility function theory is the cornerstone for most of the decision-making processes under uncertainty. In implementing this method, each problem starts with a decision tree (or a similar method) with a set of options and its probabilities assigned to each tree branch. From there other options and its possibilities are added to the end of each branch. The utility function then through using specific formula for economic good or bad (and degree of risk taking of the decision maker) will be used to calculate the utility value of each tree branch or in the other word each possible option. At the end, the decision maker will select the optimum branch based on his selected preferences.

In general, people are either risk neutral, risk averse, or risk taker and based on this personal characteristics make different decisions even if they confront the exact same conditions. So when a risk-neutral person accepts utility generated by a specific lottery game, a risk-averse person would not get into this game unless he or she is paid some money and a risk taker would be willing to pay some money to get into the same game. In Fig. 12.1, series 1, 2, and 3 in order represent the typical curves associated with risk-averse, risk-taking, and risk-neutral attitudes. If we draw a straight line parallel to the utility axis, where the line intersects with three curves are representatives of utility of sure payoff for three attitudes toward risk, while the intersection with the neutral risk line also represents the utility of accepting the alternative game as well [6].

Different literature represents different equations for calculating the utility function. The common equations that would help the decision makers make their decisions regarding an effect that higher value is considered good (first two) or bad (last two), respectively, can be described as follows:

$$U(x) = \left(1 - e^{-(X-Xmin)\,/\,\rho}\right) / \left(1 - e^{-(Xmax\,-\,Xmin)\,/\,\rho}\right)$$
$$\text{for risk-averse or risk-taker decision maker} \tag{12.1}$$

$$U(x) = (X - Xmin) / (Xmax - Xmin) \quad \text{for risk-neutral decision maker} \tag{12.2}$$

$$U(x) = \left(1 - e^{-(Xmax\,-\,X)\,/\,\rho}\right) / \left(1 - e^{-(Xmax\,-\,Xmin)\,/\rho}\right)$$
$$\text{for risk-averse or risk-taker decision maker} \tag{12.3}$$

$$U(x) = (Xmax - X) / (Xmax - Xmin) \quad \text{for risk-neutral decision maker} \tag{12.4}$$

where ρ is the risk tolerance. For an in-depth discussion about utility function and the above equations see Shultz et al. [6].

Other literatures represent the utility function as an exponential function of wealth and risk tolerance as follows:

$$U(x) = 1 - e^{-\,(X\,/\,\rho)} \tag{12.5}$$

where ρ is the risk tolerance. The higher the value of ρ, the lower risk averse is the decision maker. Finally in some literature utility function has been represented as a logarithmic function of wealth as below:

$$U(x) = \text{Ln}\,(x) \tag{12.6}$$

It should be noted that even though most of the utility functions can be modeled by either one of the above equations (of course there are other utility functions that I have not noted here as well), each person or research team or design team can develop their own risk-taking capacity curve as well. The process is as it has been shown in the next four charts. The team or individual should start with highest and lowest possible outcome of their action and assign 1 and 0 to them, respectively. They need to ask themselves for what certainty equivalence they would be convinced to accept the possible lottery 50–50 of the choices presented with 0 and 1. This certainty equivalence then will be the representative of the third point on their risk-taking curve. Following the procedure with another two point we can select multiple points on the team or individual risk-taking curve attitude. When we connect these points we will have the individual or team risk-taking attitude curve. The equation representing this curve would be the individual or team utility function (Charts 12.4, 12.5, 12.6, and 12.7).

Now and after putting together the basic concepts and thinking structure behind the two steps of decision making under uncertainty, let's try to implement these concepts in a couple of our building-related decision-making problems.

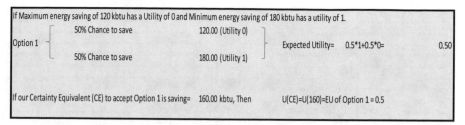

Chart 12.4 Developing own risk attitude point 1 on the curve

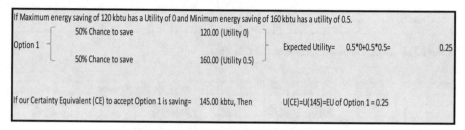

Chart 12.5 Developing own risk attitude point 2 on the curve

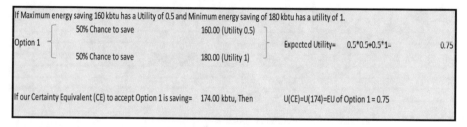

Chart 12.6 Developing own risk attitude point 3 on the curve

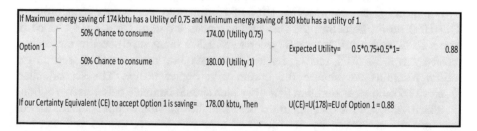

Chart 12.7 Developing own risk attitude point 4 on the curve

First let's check two conditions that in the first we are dealing with good economy (Chart 12.8) and in the second one we are dealing with bad economy (Chart 12.9) conditions. The first chart shows a condition that a decision should be made between two options in which the higher the outcome the more value the

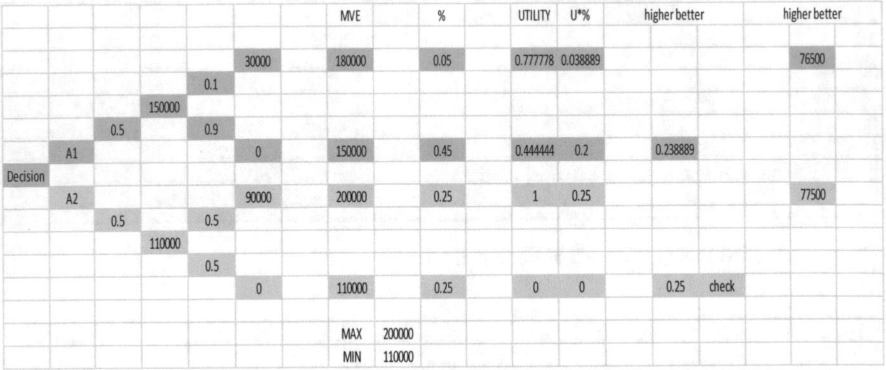

					MVE	%	UTILITY	U*%	higher better		higher better
				30000	180000	0.05	0.777778	0.038889			76500
			0.1								
		150000									
	0.5		0.9								
	A1			0	150000	0.45	0.444444	0.2	0.238889		
Decision											
	A2			90000	200000	0.25	1	0.25			77500
	0.5		0.5								
		110000									
			0.5								
				0	110000	0.25	0	0	0.25	check	
					MAX	200000					
					MIN	110000					

Chart 12.8 Decision tree of good economy function

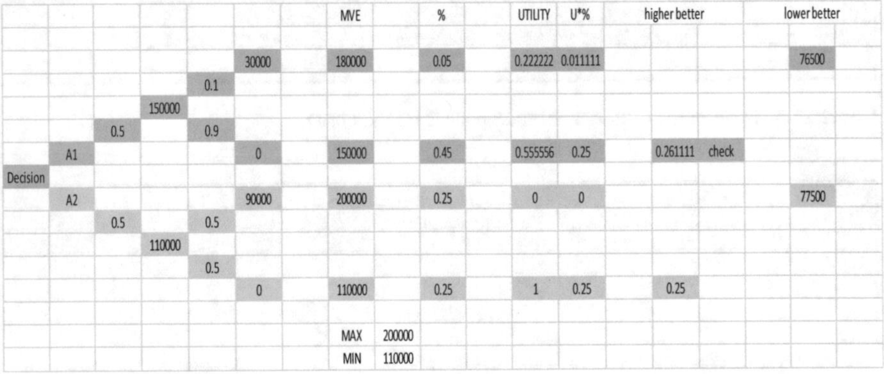

					MVE	%	UTILITY	U*%	higher better		lower better
				30000	180000	0.05	0.222222	0.011111			76500
			0.1								
		150000									
	0.5		0.9								
	A1			0	150000	0.45	0.555556	0.25	0.261111	check	
Decision											
	A2			90000	200000	0.25	0	0			77500
	0.5		0.5								
		110000									
			0.5								
				0	110000	0.25	1	0.25	0.25		
					MAX	200000					
					MIN	110000					

Chart 12.9 Decision tree of bad economy function

decision would make for us. For example assume that you are to choose between two projects that one provides $150,000 profit and the second one provides $110,000 profit. Selecting the first project opens the door for a 10 % chance of some additional work that will generate $30,000 more profit, while the second project has a 50–50 % chance for additional $90,000 more profit. Based on the utility formulas we choose the option with higher utility. The second chart (Chart 12.9) shows a condition that a decision should be made between two options in which the lower the outcome the more value the decision would make for us. For example we are asked by the owner of a project to recommend between two contractors that have provided construction cost for a specific building. Cost provided by the first contractor is $150,000 and by the second contractor is $110,000. We can go back to our file records and find that we have worked several times with each of these contractors before. Based on our records the first contractor's delivery record in average is about 90 % of times on time and only about 10 % of time up to couple of months late. The other contractor is in average about 50 % of times on time and about 50 % of time up to 6 months late. The owner immediately responses that being late on delivery schedule will cost him about $15,000 per month. Using a risk-neutral equation for bad economy will lead us to a chart similar to Chart 12.9, where selection of the first contractor will generate the best utility for this owner. For a detail explanation of a similar condition to the second example see ref. [5]

For the purpose of running an energy model with uncertainty to the best of my knowledge currently there are no commercially available software which is designed for and can thoroughly include the effects of all uncertain factors in building modeling. Here I attempt to offer a couple of simple approaches to perform such modeling with the simple available tools. In order to do our next two exercises first I have developed a simple Excel-based sheet that would take as input the deterministic hourly building cooling needs, air-handling unit airflow capacity, chilled and condenser water flow (where needed), and miscellaneous load from a traditional energy modeler run, turn them to a probabilistic input, and provide a probabilistic energy consumption output to be used for our analysis. In the second exercise I have used a random value generator and fed those to Train Trace 700 software manually before each iteration of the energy modeling. It might be a little lengthy approach, but it can provide the desired result when is done properly. Let's start with the first example here and later we would turn our attention towards the second example. Assume that based on our calculations we need a cooling system to remove equivalent of 400 t of cooling, deliver 120,000 cfm of cooling air to the building at pick load condition, and try to simulate probabilistic energy consumption (for 1 month) with three different systems: (1) water cooled chiller, cooling tower with chilled water, and condenser water pumps; (2) air-cooled chiller with chilled water pumps; and (3) package unitary direct expansion units, with the purpose of comparison and making decision on which system to select (of course the Excel sheet will be adjusted for each system, e.g., there will be no consumption for a chilled water pump in a DX system). Also assume that based on our calculations capacity of chilled water and condenser water pumps (when and if they are required) are 1000 and 1200 gpm, respectively, and a miscellaneous 5–12.5 kW load is assigned to various energy consumers of the building. (As you see I have mentioned 1-month calculations. To have a complete yearly results we need to repeat the calculations for a whole year, but here in order to simplify the process we will use 1-month analysis instead of year-around simulation). The basic source of uncertainty in performance of the HVAC equipment can be found in the level of tolerance that is allowed by the testing agencies that sets the allowed tolerance of performance for different equipment. For example assume that an equipment performance (% energy consumed to % performance delivered, e.g., KW input to tonnage delivery of a chiller) from zero to full load has been presented by Fig. 12.2. Based on the tolerance allowed by the governing testing agency if the acceptable tolerance for this chiller sets to $\pm 7.5\%$, then any chiller with the performance curve located in the margin between two upper and lower curves shown in Fig. 12.3 is representative of the possibility of the real chiller performance. Since this creates an opportunity for the manufacturers in calling the specific equipment having a higher efficiency for the curves in ~80 % of the space between the lower curve and the

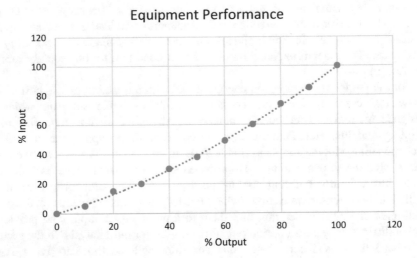

Fig. 12.2 Deterministic equipment performance curve

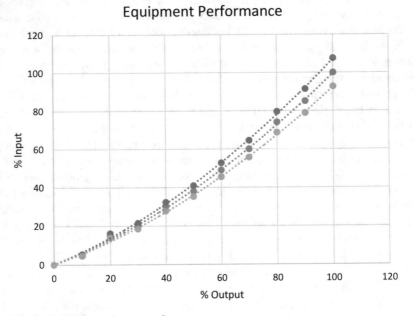

Fig. 12.3 Probabilistic equipment performance curve

middle curve, usually the equipment that are tagged with a specific performance (middle curve) in real life perform between slightly below the middle curve and the top curve. The equipment that falls immediately above the lower curve gets a tag of higher efficiency when ready to be presented to the market. Therefore in our simulation work we assume a triangle distribution for all the equipment with

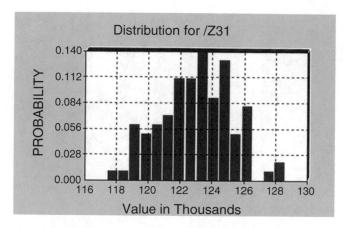

Fig. 12.4 System 1 output distribution

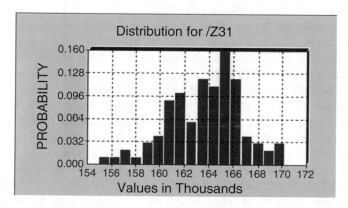

Fig. 12.5 System 2 output distribution

three possible choices of slightly below the middle curve, slightly below the upper curve, and somewhere in between. A similar situation and curve have been selected for air-handling units, and pumps with a ±5% tolerance

Then if we use a water-cooled chiller with 0.55 kW/t efficiency, an air-handling unit with 0.0015 × fan hp, a chilled water pump with 22 W/gpm, and a condenser water pump with a 19 W/gpm for system 1, an air-cooled chiller with 0.95 kW/t, with similar air-handling unit and chilled water pump to system 1 for system 2, and an air-cooled direct expansion package system with 1.1 kW/t cooling and similar air-handling side to the other two for system 3, the results of our monthly energy consumption for three different systems will be as following three figures. To perform these modeling and implementing uncertainties discussed above I have used a limited version of @risk software, which was capable of assigning random values to my targeted uncertain inputs and generate the probability distribution of the output as they are presented in Figs. 12.4, 12.5, and 12.6. For larger and more

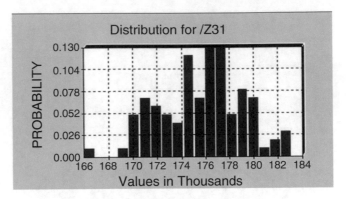

Fig. 12.6 System 3 output distribution

comprehensive modeling the full version of @risk or Phoenix Model Center software are some of the best tools to be used.

In this special case we can see that almost all of the values in Fig. 12.4 are smaller than Fig. 12.5 and almost all the values in Fig. 12.5 are smaller than Fig. 12.6. In this condition we do not need to continue our decision making since it is obvious that the most energy-consuming system is system 3 and the most energy-efficient system is system 1 under all conditions.

Now assume as if the clear winner is not so obviously recognizable and we could not simply identify the best system. Further assume that we have defined three different systems and the output distribution (energy consumption kWh/ft^2) for each can be represented as the way that I have shown it in Charts 12.10, 12.11, and 12.12. Also assume that we have the opportunity to install some onsite renewable source of electricity generation as it has shown in the middle part of each decision tree, which can reduce the energy consumption of the system by the values shown.

In developing the decision utility for each system, I have used the utility function as it is represented by Eq. 12.3 in this section. Other utility function formula such as natural logarithm methods can also be used for developing such calculations. The decision utility here is the result of sum of multiplication of utility values by their percent chances. As it can be seen under the described conditions system number 3 would provide the highest average utility value and therefore should be selected as the best answer to the problem question.

In the second exercise I developed the energy model for the multi-story office building that we have been using in the previous chapters of the book and run an energy model based on some selected values for different elements of the building in Trane Trace 700 software. The outcome obviously represents the deterministic energy consumption of the building in 1 year. In order to develop the probabilistic energy model for this building, I selected building slab U-value, roof U-value, wall U-value, glazing U-value, glazing SC (shading coefficient), supply fan hp (horse power), exhaust fan hp, air-cooled chiller efficiency (kw per ton), chilled water pump gpm (gallons per minute), and condenser water pump (gallons per minute) as

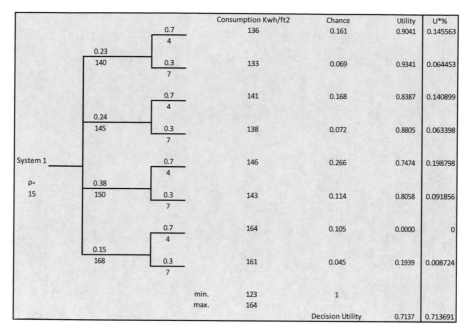

			Consumption Kwh/ft2	Chance	Utility	U*%
		0.7	136	0.161	0.9041	0.145563
		4				
	0.23					
	140	0.3	133	0.069	0.9341	0.064453
		7				
		0.7	141	0.168	0.8387	0.140899
		4				
	0.24					
	145	0.3	138	0.072	0.8805	0.063398
		7				
System 1		0.7	146	0.266	0.7474	0.198798
		4				
ρ=	0.38					
15	150	0.3	143	0.114	0.8058	0.091856
		7				
		0.7	164	0.105	0.0000	0
		4				
	0.15					
	168	0.3	161	0.045	0.1939	0.008724
		7				
	min.		123	1		
	max.		164			
				Decision Utility	0.7137	0.713691

Chart 12.10 System 1 decision utility

			Consumption Kwh/ft2	Chance	Utility	U*%
		0.7	126	0.091	0.9846	0.089599
		4				
	0.13					
	130	0.3	123	0.039	1.0000	0.039
		7				
		0.7	136	0.175	0.9041	0.158221
		4				
	0.25					
	140	0.3	133	0.075	0.9341	0.070058
		7				
System 2		0.7	156	0.357	0.4421	0.157817
		4				
ρ=	0.51					
15	160	0.3	153	0.153	0.5558	0.085037
		7				
		0.7	161	0.077	0.1939	0.014927
		4				
	0.11					
	165	0.3	158	0.033	0.3526	0.011635
		7				
	min.		123	1		
	max.		164			
				Decision Utility	0.6263	0.626293

Chart 12.11 System 2 decision utility

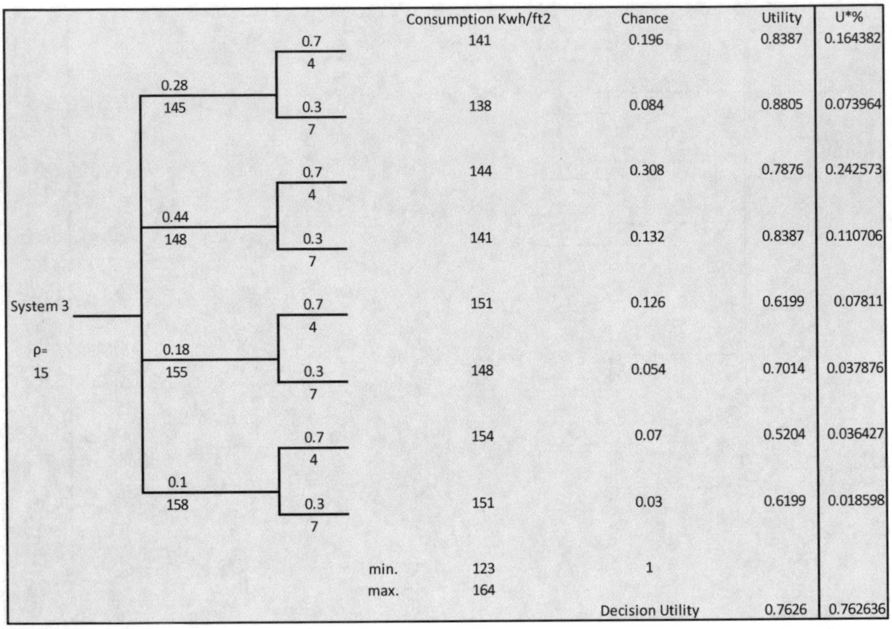

			Consumption Kwh/ft2	Chance	Utility	U*%
		0.7	141	0.196	0.8387	0.164382
		4				
	0.28					
	145	0.3	138	0.084	0.8805	0.073964
		7				
		0.7	144	0.308	0.7876	0.242573
		4				
	0.44					
	148	0.3	141	0.132	0.8387	0.110706
		7				
System 3		0.7	151	0.126	0.6199	0.07811
		4				
ρ=	0.18					
15	155	0.3	148	0.054	0.7014	0.037876
		7				
		0.7	154	0.07	0.5204	0.036427
		4				
	0.1					
	158	0.3	151	0.03	0.6199	0.018598
		7				
		min.	123	1		
		max.	164			
				Decision Utility	0.7626	0.762636

Chart 12.12   System 3 decision utility

uncertain inputs each with a +-5% tolerance in their selected values. Separately I generated 20 sets (it would be more reliable to use about 80 iterations, but to the benefit of time I just made 20 iterations) of random selection within each element upper and lower limits and then fed them one by one to the software and run the simulation 20 times to come up with the probability distribution of the building and source energy consumption for this building; the results are represented in Figs. 12.7 and 12.8. Similar to what we did for the first exercise, information that is gathered from the below probability distribution curves can be utilized in proper decision trees to help the decision makers to make appropriate decisions regarding this building energy consumption. To produce the following two curves I used normal curve calculator presented by developers noted in reference [7]. Obviously it makes more sense for the interested designers or even the modeling software developer to add or include an algorithm capable of automatically iterating each single energy model without the need for manually repeating each iteration.

Bayesian Rules

Subjective (personalistic) probability is an old notion. As early as in the Arsconjectandi (1713) by Jacques Bernoulli (1654–1705, an uncle of Nicolas and Daniel) probability was defined as a degree of confidence that may be different with different persons. The use of

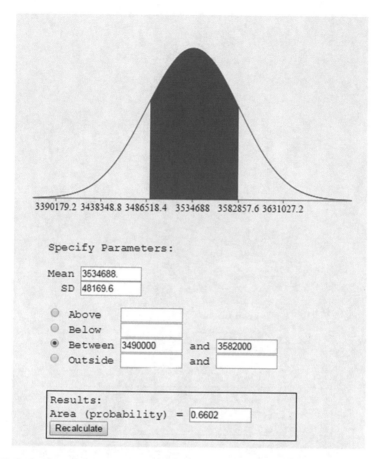

Fig. 12.7 Probability distribution of building energy consumption (Btu/year) with 66 % percentile probability highlighted

subjective probabilities in expected utility theory, was, however, first developed by Frank Ramsey in the 1930's. Expected utility theory with both subjective utilities and subjective probabilities is commonly called *Bayesian decision theory*, or Bayesianism. [8]

Up to this point, every discussion that we had in the previous section in order to develop our utility function was based on objective utility assigning and probabilities. As it was explained and utilized earlier, objective probabilities are those probabilities that could be measured or calculated using empirical and actually recorded observations and evidences, such as the result of the energy modeling in calculation of energy consumption of the building in previous section. When such hard evidence and calculations are not available regarding a specific item, sometimes we may suffice on pursuing our approach by targeting another type of probability, which is known to be the subjective probability. As it can be found in different literatures a subjective probability can be defined as the result of judgement and therefore the opinion of the

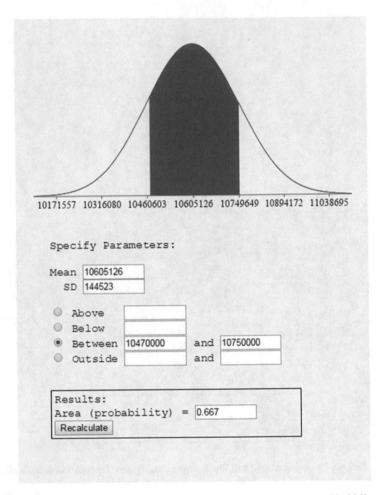

Fig. 12.8 Probability distribution of source energy consumption (Btu/year) with 66 % percentile probability highlighted

decision maker about possibility of an outcome, and it is not required to be based on any formal calculation. When Von Neumann developed the utility function, he based the utility function on objective probabilities. As I said earlier a few years after development of the Von Neumann objective method, Savage showed the subjective probability also can be used in utility function development as long as it is built upon proper evidence. His work was based on and in some form complimentary to Bayesian Theory which was developed for the first time by Thomas Bayes (1701–1761). Thomas Bayes's Bayesian theory defines the original probability of some events based on the subjective belief of the decision maker first, and then and upon occurrence of new evidences attempts to update the probability of the events.

The main and basic Bayesian equation is

$$P\left(A|B\right) = P\left(B|A\right) \times P\left(A\right)/P\left(B\right) \qquad (12.7)$$

which can be read as the probability of event "A" (also called hypothesis) given event "B" (also called evidence) has happened is equal to probability of event "B" given the event "A" has happened multiplied by the prior probability of event "A," divided by the prior probability of event "B." This equation can be developed to two more general forms for single evidence with multiple hypothesis (Eq. 12.8) and multiple evidence with multiple hypothesis (Eq. 12.9): (see [9])

$$P\left(Ai|B\right) = P\left(B|Ai\right) \times P(Ai)/\Sigma_{k=1 \text{ to } m}P\left(B|A_K\right) \times P\left(A_K\right) \qquad (12.8)$$

And:

$$
\begin{aligned}
P\left(Ai|B_1B_2B_3...B_n\right) = {} & P\left(B_1|Ai\right) \times P\left(B_2|Ai\right) \times P\left(B_3|Ai\right) \\
& \times \cdots \times P\left(B_n|Ai\right) \times P\left(Ai\right)/\Sigma_{k=1 \text{ to } m}P\left(|B_1|A_K\right) \times P\left(|B_2|A_K\right) \\
& \times P\left(|B_3|A_K\right) \times \cdots \times P\left(|B_n|A_K\right) \times P\left(A_K\right)
\end{aligned}
$$

$$(12.9)$$

In twentieth century and following the works of Sir Harold Jeffrey (1891–1989), Abraham Wald (1902–1950) and Leonard Savage (1917–1971) and their works "Theory of Probability," "Statistical Decision Functions," and "Foundations of Statistics," respectively, Bayesian method reemerged and two interpretations of objective and subjective probability became the focus of the discussion among scientists. The main difference between these two points of view was based on the different opinion on how to construct the prior probabilities. The scientists that leaned towards the objective probability view focused on crisp model, data, requirements of rationality, and consistency which everyone (community of scientists) should believe based on the available knowledge and data, and those who were leaned towards the subjective probability view focused on quantifying the probability based on "decision maker's own belief" as long as it proves to be within the constraints of rationality and coherency. In previous section we dealt with objective side of the utility function, so here let's direct our focus towards explaining how to use the subjective point of view in Bayesian probability which can be developed based on an expert belief in order to select an option among different possibilities.

In order to explain this let's assume the process in hand is to select the best choice among three equipment manufacturers who are the (sole or at least main) providers of the equipment that we want to specify for our design project. This process is a subjective approach since we cannot use tools such as energy modeling or well defined equations such as what were the cases used in utility function that we described earlier in this section. The criteria can be explained with more clarity as selecting relative utility that we can get among manufacturer $A1$, $A2$, and $A3$ and based on three criteria of last longer equipment, less complicated system which reduces the need for highly trained maintenance personnel and providing access to more knowledgeable representatives.

The prior probability of the manufacturers' relative utility (should sums up to 1) comes from the belief of the decision maker. Let's assume based on our experience with these three manufacturers our prior probability of which manufacturer provides the best equipment is 0.4, 0.3, and 0.3 for manufactures $A1$, $A2$, and $A3$, respectively. Also assume our expert can project 0.3, 0.7 and 0.6 (Does not need to sum up to 1) values to the possibility of evidence 1, 2, or 3 observed giving the manufacturer $A1$ is selected, 0.6, 0.5, 0.1 values to the possibility of evidence 1, 2, or 3 observed giving the manufacturer $A2$ is selected, and 0.4, 08, 0.6 values to the possibility of evidence 1, 2, or 3 observed giving the manufacturer $A3$ is selected.

Using these values and Eq. 12.8, one evidence after the other, we now can define the possibility of selecting (which one is better) manufacturer $A1$, $A2$, and $A3$ giving the values that has been developed based on our expert belief (see Charts 12.13, 12.14, 12.15, and 12.16 below).

The Bayesian approach requires us to search for new evidence and update our prior beliefs according to these new evidences. To do so we need to investigate our available data and also communicate with other similar companies and resources that may have recent experience with either of these three manufacturers, and upon new evidence renew our calculations and the level of our belief on which manufacturer is the best to be selected.

Chart 12.13 Expert's believed probabilities and hypothesis of the problem

Probability	Hypothesis		
	i=1	i=2	i=3
P(Ai)	0.4	0.3	0.3
P (B1|Ai)	0.3	0.6	0.4
P (B2|Ai)	0.7	0.5	0.8
P (B3|Ai)	0.6	0.1	0.6

$$P (A1|B1) = (0.3 \times 0.4) / (0.3 \times 0.4 + 0.6 \times 0.3 + 0.4 \times 0.3) =\quad 0.29$$
$$P (A2|B1) = (0.6 \times 0.3) / (0.3 \times 0.4 + 0.6 \times 0.3 + 0.4 \times 0.3) =\quad 0.43$$
$$P (A3|B1) = (0.4 \times 0.3) / (0.3 \times 0.4 + 0.6 \times 0.3 + 0.4 \times 0.3) =\quad 0.29$$

Chart 12.14 Probability of betterment of $A1$, $A2$, and $A3$ given realization of evidence of $B1$

$$P (A1|B1B2) = (0.3 \times 0.4 \times 0.7) / (0.3 \times 0.4 \times 0.7 + 0.6 \times 0.3 \times 0.5 + 0.4 \times 0.3 \times 0.8) =\quad 0.31$$
$$P (A2|B1B2) = (0.3 \times 0.5 \times 0.6) / (0.3 \times 0.4 \times 0.7 + 0.6 \times 0.3 \times 0.5 + 0.4 \times 0.3 \times 0.8) =\quad 0.33$$
$$P (A3|B1B2) = (0.4 \times 0.3 \times 0.8) / (0.3 \times 0.4 \times 0.7 + 0.6 \times 0.3 \times 0.5 + 0.4 \times 0.3 \times 0.8) =\quad 0.36$$

Chart 12.15 Probability of betterment of $A1$, $A2$, and $A3$ given realization of evidences of $B1$ and $B2$

$P(A1|B1B2B3) = (0.3 \times 0.4 \times 0.7 \times 0.6) / (0.3 \times 0.4 \times 0.7 \times 0.6 + 0.6 \times 0.3 \times 0.5 \times 0.1 + 0.4 \times 0.3 \times 0.8 \times 0.2) =$ 0.43

$P(A2|B1B2B3) = (0.3 \times 0.5 \times 0.6 \times 0.1) / (0.3 \times 0.4 \times 0.7 \times 0.6 + 0.6 \times 0.3 \times 0.5 \times 0.1 + 0.4 \times 0.3 \times 0.8 \times 0.2) =$ 0.08

$P(A3|B1B2B3) = (0.4 \times 0.3 \times 0.8 \times 0.6) / (0.3 \times 0.4 \times 0.7 \times 0.6 + 0.6 \times 0.3 \times 0.5 \times 0.1 + 0.4 \times 0.3 \times 0.8 \times 0.6) =$ 0.49

Chart 12.16 Probability of betterment of $A1$, $A2$, and $A3$ given realization of evidences of $B1$, $B2$, and $B3$

For a detail and comprehensive discussion about basics of Bayesian statistics, modeling, and analysis refer to Negnevitsky [9] and Sebastisni [10].

References

1. Swiler, L. P., Paez, T. L., & Mayes, R. L. (2009). Epistemic uncertainty quantification tutorial. *Proceedings of the IMAC XXVII Conference and Exposition on Structural Dynamics.*
2. Kiureghian, A. D. (2007). *Aleatory or epistemic? Does it matter?* Special Workshop on Risk Acceptance and Risk Communication March 26–27, 2007, Stanford University, University of California, Berkeley.
3. Khazaii, J. (2014). *Energy efficient HVAC design, an essential guide for sustainable building.* New York, NY: Springer.
4. American Society of Heating, Refrigerating and Air Conditioning Engineers. (2013). *ASHRAE 90.1 Standard: Energy Standard for buildings except low-rise residential buildings.* Atlanta, GA: American Society of Heating, Refrigerating and Air Conditioning Engineers.
5. Hulett & Associates web site; use decision Trees to make important project decision; http://www.projectrisk.com/decision_tree_for_important_project_decisions.html
6. Schultz, M. T., Mitchell, K. N., & Harper, B. K. (2010). *Decision making under uncertainty.* ERDC TR-10-12. U.S. Army Corps of Engineers.
7. Online statistics education: An interactive multimedia course of study; Developed by Rice University (Lead Developer) and University of Houston Clear Lake and Tufts University. http://onlinestatbook.com/2/calculators/normal_dist.html
8. Hansson, S. O. (2005). *Decision theory, a brief introduction.* Stockholm: Royal Institute of Technology.
9. Negnevitsky, M. (2005). *Artificial intelligence, a guide to intelligent systems.* Harlow: Pearson Education.
10. Sebastiani, P. (2010) *A tutorial on probability theory.* Department of Mathematics and Statistics; University of Massachusetts at Amherst. http://www.sci.utah.edu/~gerig/CS6640-F2010/prob-tut.pdf

Chapter 13
Agent-Based Modeling

Abstract Agent-based modeling is a computational simulating method which uses autonomous agents that live, move, and interact with each other on a virtual environment (world). In addition to capability of interacting with each other, the agents are capable of affecting and changing the world as well. Agent-based modeling is a great tool for researchers to define the initial properties of agents and environment and also rules of change and interaction and observe what would emerge from these interactions after a period of time is passed.

Keywords Agent-based modeling • Agent • World • Discrete events • Feedback loops • Dynamic systems • Intelligent behavior • Occupants' dynamic behavior • Autonomous behavior • Process-oriented perspectives • Microelectromechanical system

In the world of models, analytical (static) models are those that can be defined based on the fact of being strictly depending on their input and hard coded mathematical formulas. In other words the solution to the formula is used to describe the state of the system. The underperformance with a static model however is that when the question subject becomes more complicated the solution may become very hard to find. To the contrary simulation (dynamic) models have the advantage of providing an environment under which based on a set of rules we may be able to predict how the state of a complicated situation may change over time with more simplicity. Traditionally the main paradigms in simulation (dynamic) modeling has been categorized to system dynamics (in which model tended to deal with continuous processing through mapping and modeling stages along with nesting feedback loops inside the system in order to increase the validity of the results and it was originally developed to improve understanding of industrial processes for corporate managers in 1950s), discrete events (in which inputs entered the system and passed from one function to another in discrete time intervals with no change in the system between the consecutive events), and dynamic systems (which was mostly applicable in modeling and designing the time varying behavior of physical systems with particular properties, and used mathematical equations that describe these time-based systems). For a detail discussion on this refer to ref. [27]. A newer paradigm that has recently emerged in the dynamic modeling realm is called agent-based modeling.

© Springer International Publishing Switzerland 2016
J. Khazaii, *Advanced Decision Making for HVAC Engineers*,
DOI 10.1007/978-3-319-33328-1_13

Agent-based modeling is a computational simulating method which uses autonomous agents that live, move and interact with each other on a virtual environment (world). In addition to capability of interacting with each other, the agents are capable of affecting and changing the world as well. Agent-based modeling is a great tool for researchers to define the initial properties of agents and environment and also rules of change and interaction and observe what would emerge from these interactions after a period of time is passed. This approach not only creates the opportunity to analyze some dynamic phenomena which we were not able to evaluate before (such as emergent, self-organizing phenomena, complex organizational models, and realistic traffic simulations), but also changes the old and usually dominant top-down and holistic modeling approach in which the whole phenomena was modeled with mathematical equations and rules only, without capability of evaluating the possible changes that can be enforced to the process as the result of some unpredicted bottom up action of building blocks of the phenomena (agents). Even though the behavior of the agents sometimes are represented by mathematical equations, but most of the times these behavior are following the if–then type of logical rules. This makes this type of modeling more flexible and therefore susceptible to create more realistic results. Kugel [1] depicts the advantages of agent-based modeling in comparison to the other traditional simulations as its not imposing restriction for design, no need for homogenous population, space or restrictedly uniform interactions, ability of representing intelligent behavior, generating macro-level phenomena based on micro-level behavior and freedom of the designer to include everything he thinks is necessary in the model.

Different researchers have evaluated agent-based modeling processes against the methods of deduction and induction in different literatures. As Axelrod [2] explains "The purpose of induction is to find patterns in data and that of deduction is to find consequences of assumptions, the purpose of agent-based modeling is to aid intuition" and, as Billari et al. [3] indicate: "As with deduction, agent-based modeling starts with assumptions. However, unlike deduction, it does not prove theorems. The simulated data of agent-based models can be analyzed inductively, even though the data are not from the real world as in case of induction." Macal and North [4] in regards to agent-based modeling and simulation say, "ABMS promises to have far-reaching effects on the way that businesses use computers to support decision-making and researchers use electronic laboratories to do research. Some have gone so far as to contend that ABMS is a new way of doing science."

Macal and North [5] states "A combination of several synergistic factors is moving ABMS forward rapidly. These factors include the continuing development of specialized agent-based modeling methods and toolkits, the widespread application of agent-based modeling, the mounting collective experience of the agent-based modeling community, the recognition that behavior is an important missing element in existing models, the increasing availability of micro-data to support agent-based models, and advances in computer performance."

Kügl [1] defines several areas that using an agent-based modeling can be more advantageous than other simulation modeling methods, such as systems where decision making happens on different levels of aggregation, multi-level systems that require observation capabilities beyond one level, systems that draw their dynamics from flexible and local interaction, and system which transient dynamics should be analyzed.

Many social, behavioral, and biological concepts have been researched, analyzed, and advanced by using agent-based modeling method. Subjects such as people behavior when facing decisions that involve considering other people decisions, animal flocking, and epidemic disease control are to name just a few of them.

In engineering related research, Treado and Delgoshaei [6] in their paper through an application of a system designed with on-site power generation, renewable energy sources, thermal storage, and grid integration represented "how agent-based methods could be incorporated into the operation of building control systems, and thereby improving building energy performance, comfort and utility."

Zeng et al. [7] in their paper represented "the implementation of a multi agent system (MAS) for distributed renewable energy (RE) management of a set of eco-building units." They through agent-based modeling demonstrated that a hybrid renewable energy power system can balance the power supply and load demand, while meeting the system requirements. Davidsoon and Boman [8] in their paper suggested a multi-agent model to improve the control design for intelligent buildings, and proposed an approach that would help to optimize the tradeoff between energy consumption and customer satisfaction.

Von Breemen [9] presented a method that "helps the designer to solve complex control problems, by offering concepts to structure the problem and to organize the solution. The design method encourages to develop local solutions and to reason about their dependencies. It offers different coordination mechanisms to deal with these dependencies." MacLeod and Stothert [10] used a society of communicating agents to design a system capable of controlling the ambient temperature in deep-level mines. MacKenzie [11] used an agent-based model which uses a set of agents which are operating at the same time for controlling moving robots. Lygeros et al. [12] presented an agent-based solution for applications that demands efficient use of scarce resources such as aircraft request for runway space.

One of the least developed sectors in building energy modeling is evaluation of the effects of the occupants' dynamic behavior in outcome of the energy modeling. That is a major missing component from energy modeling practice in spite of the fact that research has shown generally equally importance of human behavior which in some cases can be even more influential on energy consumption of building than system behavior (See Masoso and Grobler [13] research which indicated auditing a few buildings in hot and dry climate showed unexpected higher energy consumption during non-operating hours than operating hours due to the occupant behavior.). Researchers have been trying to overcome this shortfall by offering interesting methods in the past few years. In one of these attempts, Lee [14] based on behavioral theory of "reason action model" has been successful to develop an interaction model of the occupants and the building comfort/energy consumption controls to quantify this behavior into energy modeling by coupling EnergyPlus and agent-based modeling platforms. He represented "The agent-based modeling approach for simulating occupant behaviors has shown that uncertainties of occupant behaviors can be accounted for and simulated, lending the potential to augment existing simulation methods."[14]

Liaoa and Barooaha in 2010 pointed out "Currently the most popular method of incorporating information about time-variations in occupancy to energy simulations is through schedules and diversity factors (Abushakra et al.) [15]. However, these methods are not meant to capture occupancy dynamics. Diversity factors, for example, are used to provide a correction to mean heat gain calculations from occupancy schedules so that the mean heat gain so computed are more representative of the true mean value. A model of occupancy dynamics can provide more realistic sample time-traces of occupancy over time from which, peaks, means and variances of occupancy driven heat gains can be computed." [16]

Erickson et al. [17] used an agent-based modeling solution to show in buildings knowing the occupancy and usage patterns will result in significantly higher energy savings in comparison with the current assumption of fixed occupancy and usage.

Some of the other researches done in the field that can be noted here are: "Agent-based modeling has been used to simulate the diffusion of energy saving ideas through social networks within a building (Chen, Taylor and Wei) [18], predict user-controlled plug loads (Zhang, Siebers and Aickelin) [19] and it has allowed realistic representations of occupants' adaptive responses to changing comfort conditions (Andrews et al.) [20]." [21]

All these researches show that the agent-based modeling has the promise of having enough strength to be used as a very valuable tool in finding solutions for complicated HVAC and energy related problems. So let us discuss the method in a little more depth. As it can be understood from the agent-based model title an important element of this type of modeling is an "Agent." An agent can be assumed to be a human, an animal, an object a particle, etc. An agent's most important characteristic is its autonomous behavior. This means the agent can (make decision and) function completely independently. As Macal and North [4] explain as a general rule each agent has its associated attributes, behavioral rules, memory, resources, decision making sophistication and rules to modify behavioral rules. Jenings et al. [22] explain the minimum attributes required for an entity to be called an agent are situatedness (being in an environment to be able to receive input and change the environment), autonomous (having control without direct human intervention), responsive (being capable of timely response), pro-active (being capable of initiative actions), and social-ability (being able to interact—this is the great value of this type of modeling that allows the modeler to let the independent agents to interact with each other and also with the "world" in order to check the possible emergence in the state of the agents and the world as the result of the interactions). Wooldrige [23] also names mobility, reasoning model and learning as possibilities that could be but are not essential attribute of an agent.

As it is appeared from the previous paragraph another important element in agent-based modeling is the "world." The world is the environment that the agents can live in, move over, use, and change it if it is required. The world itself can have its own associated attributes, behavioral rules, memory, and resources. Gilbert [24] states environment may be an entirely neutral medium with little or no effect on the agents, or in other models, the environment may be as carefully crafted as the agents themselves.

In order to develop and implement a proper agent-based model we should define the agents and their attributes, define the environment and its attributes, define the procedures of updating the attributes of the agents and the environment and methods of interacting among them. Klügl [1] summarizes the best practices for performing successful agent-based modeling as (1) defining the problem clearly from the domain side, (2) gathering appropriate and relevant data and literatures, (3) having access to external expert for validation and review of the model, (4) common agreement among all the participants on suitability of the agent-based modeling for the problem in hand, (5) close cooperation between the domain and modeling experts if they are different people, (6) suitable model concept and design, (7) adoption to the agent and process-oriented perspectives, (8) minimizing the uncertainties in the model, (9) efficient technical design, (10) proper software, and (11) proper documentation throughout the design process. There are many software available for performing agent-based modeling, such as Brahms, Starlogo, Ascape, and Netlogo. Netlogo is freely available for download at "https://ccl. northwestern.edu/netlogo/" and has a great tutorial and library of models at "https://ccl.northwestern.edu/netlogo/models/index.cgi". By reading through these models and software tutorials one can develop simple, yet strong models. For a simple experiment let us choose Thermostat model developed by Wilensky [25, 26]. The model provides a semipermeable border that allows some heat leaves the border (room walls). A thermometer check the room temperature and if the temperature is below the set-point a heater starts heating the room until the thermometer sense the temperature is reached the set-point, and the cycle continues while the temperature chart shows temperature move above and below the set-point temperature. That is a simple model of how in the real world Thermostats and heaters operate. When we set the temperature set-point (say at 67 F), level of energy that is released from the heater and also the level of permeability of the walls and then run the program we can see how agents (representatives of temperature) move inside the environment (space inside the room) and interact with another agent (thermometer) and in different intervals causes the third agent (heater) to operate in order to keep the set-point of the thermometer at a specific level. Temperature oscillates around the set-point similar to what happens in real life models which is due to over-shooting or under-shooting of the heater in order to match the thermometer set-point. In order to make the experience even more interesting let us revise the model code and add a second thermometer to the room, keep all the other parameters same as the original experiment and run the program one more time. We will have a new temperature curve. In order to make the experiment even more exciting we can use the "Behavior Space" command and run each experiment multiple time. If we collect the results of all the multiple runs for each condition we can see that the average temperature in second option (two thermometers) is

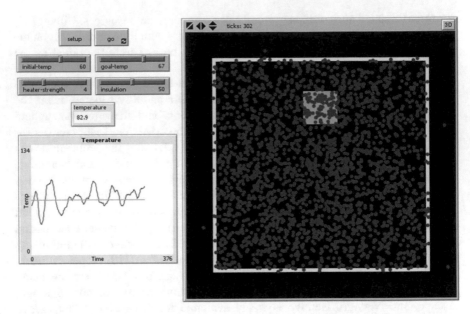

Fig. 13.1 Courtesy of [25, 26]

much closer to the set-point (67 F) than the average temperature in first option. Also we can observe that with second option the minimum and maximum off-shoots are smaller than the same measure for the first option. Such experiment can show that for each space installing two thermometers in different locations of the room causes a better control that when we install only one thermometer for that space. Of course the results shall be experienced in real systems to confirm the finding from the program (Fig. 13.1).

If we continue the experiment with additional thermometers in the room and in each case we reach the same conclusion as the above experiment we may reach to the conclusion that the more thermometers in a space the tighter the control of the temperature and as a result the higher the level of comfort for occupants could be achievable. Such experiment if is confirmed in real life testing, could justify the use of multiple temperature sensors (microelectromechanical system (MEMS) sensors) in the building, instead of the current traditional single temperature sensor (Fig. 13.2.).

Klügl [1] defines issues that make developing a good agent-based model program challenging. Some of these issues are, the link between micro- and macro-side of the program, emergence and non-linearity (emergence of some qualitative new pattern that are not reducible by current rules), brittleness and sensitivity (sometimes slightly different initial condition results in completely different outcomes), tuning micro-level and falsification (tuning the behavior of agents to reach a desired predefined outcome), level of detail and number of assumptions, size and scalability, and understanding the types of problems that are not suitable to be solved by agent-based modeling.

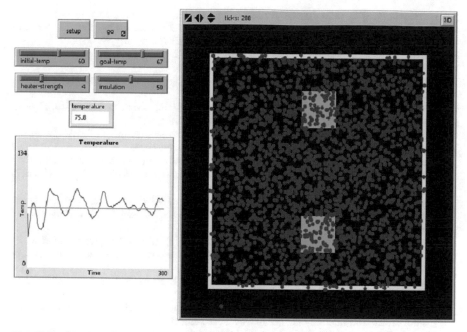

Fig. 13.2 Courtesy of [25, 26]

Agent based modeling is a great tool that can help researchers reach valuable conclusions in cases where actual experiment could prove to be very expensive and sometimes hard to perform.

References

1. Klügl, F. *Agent-Based Simulation Engineering*. http://www.oru.se/PageFiles/63259/mas_pdfs/kluegl_habilitation.pdf
2. Axelrod, R. (1997). *The complexity of cooperation. Agent-based models of competition and collaboration*. Princeton: Princeton University Press, Princeton, New Jersey.
3. Billari, F. C., Fent, T., Prskawetz, A., Scheffran, J. (2006). Agent-based computational modelling: An introduction. http://todd.bendor.org/upload/fulltext_001.pdf
4. Macal, C. M., North, M. J. (2006). Tutorial on agent-based modeling and simulation Part-2: How to model with agents.
5. Macal, C. M., & North, M. J. (2010). 2010; Tutorial on agent-based modelling and simulation. *Journal of Simulation, 4*, 151–162.
6. Treado, S., Delgoshaei P., (2010) Agent-Based Approaches for Adaptive Building HVAC System Control, Purdue university, Purdue e-pubs.
7. Zeng, Jun, W. U. Jie, Lie Jun-feng, Gao La-mei, and L. I. Min. (2008). An agent-based approach to renewable energy management in eco-building. *International Conference on Sustainable Energy Technologies*: 46–50.

8. Davidsson, P., Boman, M. (2000). A multi-agent system for controlling intelligent buildings. Department of computer science, University of Karls Krona/Ronneby, Soft Center, and Department of Computer and Systems Sciences, Stockholm University, Sweden.
9. Von Breemen, A. J. N. (2001). Agent-based multi-controller systems. A design framework for complex control problems.
10. MacLeod, I. M. and Stothert, A. (1998). Distributed intelligent control for a mine refrigeration system. *IEEE Control Systems*, 31–38.
11. Lygeros, J., Godbole, D. N., & Sastry, S.(1997). A design framework for hierarchical, hybrid control. California PATH Research Report, UCB-ITS-PRR-97-24.
12. MacKenzie, D. C. (1996). *A Design Methodology for the Configuration of Behavior-Based Mobile Robots*. PhD thesis, Georgia Institute of Technology, 1996.
13. Masoso, O. T., & Grobler, L. J. (2010). The dark side of occupants' behaviour on building energy use. *Energy and Buildings, 42*(2), 173–177, School of Mechanical and Materials Engineering North-West University, Potchefstroom.
14. Lee, Y. S. (2013). *Modeling multiple occupant behaviors in buildings for increased simulation accuracy: An agent-based modeling approach*; PhD Thesis. Philadelphia, PA: University of Pennsylvania.
15. Abushakra, B., Sreshthaputra, A., Haberl, J., & Claridge, D. (2001). *Compilation of diversity factors and schedules for energy and cooling load calculations*. Technical report, Energy Systems Laboratory, Texas A and M University.
16. Liaoa, C., Lina, Y., Barooaha, P. (2010). Agent-based and graphical modeling of building occupancy
17. Erickson, V. L, Lin, Y., Kamthe, A., Brahme, R. (2010). Energy efficient building environment control strategies using real-time occupancy measurements.
18. Chen, J., Taylor, J. E., & Wei, H. H. (2012). Modeling building occupant network energy consumption decision-making: The interplay between network structure and conservation. *Energy and Buildings, 47*, 515–524.
19. Zhang, T., Siebers, P. O., & Aickelin, U. (2011). Modelling electricity consumption in office buildings: An agent based approach. School of computer science, university of Nottingham, Nottingham, U.K.
20. Andrews, C. J., Yi, D., Krogmann, U., Senick, J. A., & Wener, R. E. (2011). Designing buildings for real occupants: An agent-based approach. *IEEE Transactions on Systems, Man, and Cybernetics--Part A: Systems and Humans, 41*(6), 1077–1091.
21. Figueroa, M., Putra, H. C., & Andrews, C. J. (2014). Preliminary Report: Incorporating Information on Occupant Behavior into Building Energy Models. Prepared by the Center for Green Building at Rutgers University for the Energy Efficient Buildings Hub, Philadelphia, PA.
22. Jennings, N. R., Sycara, K., & Wooldridge, M. (1998). A roadmap of agent research and development. *Autonomous Agents and Multi-Agent Systems, 1*, 275–306.
23. Wooldridge, M. (1997); Agent-based computing. *Baltzar Journals*
24. Gilbert, G. N. (2008). Agent-based models, university of surrey, Guildford, U.K.
25. Wilensky, U. (1998). *NetLogo Thermostat model*. http://ccl.northwestern.edu/netlogo/models/ Thermostat. Center for Connected Learning and Computer-Based Modeling, Northwestern University, Evanston, IL.
26. Wilensky, U. (1998). *NetLogo*. http://ccl.northwestern.edu/netlogo/. Center for Connected Learning and Computer-Based Modeling, Northwestern University, Evanston, IL.
27. Borshchev, A., Filippov, A., From system dynamics and discrete event to practical agent based modeling: Reasons, Techniques Tools. (XJ Technologies and St. Petersburg Technical University)

Chapter 14
Artificial Neural Network

Abstract One of the most intriguing science stories that I have heard is the Story of Science by Michael Mosely as a BBC TV production under the same title. In a part of the story he recognizes the lifelong collected data about the locations and movements of stars and planets in the sky by a Danish astronomer with the name of Tycho Brahe in late 1500s and early stages of 1600s as the main source of the great German mathematician Kepler's mathematical findings about the planetary movement of the known universe of their time. The story more evolves as the great Italian renaissance-man Galileo invents the first telescope and verifies the findings of Tycho Brahe and Kepler with a much closer look into the sky. Finally Newton the British father of the industrial ages put an exclamation point at the end of these findings by developing his famous planetary gravitation law which has been the major source for many of the scientific findings in its following 350 years. Such stories not only show the historic importance of collection of good raw data, even though at the time people may do not recognize how to make sense of that, but also prevails the need for some appropriate tools for drawing patterns and analyzing this data for developing rules and laws to advance and ease our understanding of any phenomenon.

Keywords Neural system • Neuron • Axon • Dendrite • Artificial neural network algorithm • Sigmoid function • Tangent hyperbolic function • Back-propagation • Threshold • Bias

One of the most intriguing science stories that I have heard is the Story of Science by Michael Mosely as a BBC TV production under the same title. In a part of the story he recognizes the lifelong collected data about the locations and movements of stars and planets in the sky by a Danish astronomer with the name of Tycho Brahe in late 1500s and early stages of 1600s as the main source of the great German mathematician Kepler's mathematical findings about the planetary movement of the known universe of their time. The story more evolves as the great Italian renaissance-man Galileo invents the first telescope and verifies the findings of Tycho Brahe and Kepler with a much closer look into the sky. Finally Newton the British father of the industrial ages put an exclamation point at the end of these findings by developing his famous planetary gravitation law which has been the major source for many of the scientific findings in its following 350 years.

Such stories not only show the historic importance of collection of good raw data, even though at the time people may do not recognize how to make sense of that, but also prevails the need for some appropriate tools for drawing patterns and analyzing this data for developing rules and laws to advance and ease our understanding of any phenomenon. We are living in a data rich world. The advent of the computers in the past two decades has made it possible to gather a plethora of raw data for any business or occasion. Alongside of these new opportunities for gathering copious data has emerged methods and techniques that would be able to help the decision makers make better sense of this large available raw data. In the following paragraphs of this section I will discuss one of the most important pattern recognition methods briefly.

Neural system is a very complicated part of the human body that communicates (receives and sends) signals from and to different parts of its body. Generally nervous system is divided into central and peripheral nervous system where brain and spinal cord create the central and nerves create the peripheral nervous systems. The main cell in a nervous system is Neuron (nerve cell). Neurons exist in three types, sensory neurons that sense the outside world, interneurons that carry the signal from the sensory neurons to brain and back to last group of neurons that are motor neurons and are responsible for sending signals to muscles and glands for reaction to what has been sensed by the sensory neurons. Neuron has a central part called soma which itself contains a nucleus. A single slender and relatively long fiber called axon exits the soma and extends away from cell. Axon's duty is to conduct an electrochemical impulse away from the cell in order to make the impulse accessible for the other downstream targeted nerve cells.

At the tail-end of the axon there are axon's terminals that are capable of transmitting the same pulse to multiple receivers without any degradation of the plus as it chooses these multiple paths. Therefore multiple receivers would be capable of receiving the same pulse strength as it was traveling through the upstream axon. On the receiving end of the neuron there are dendrites which have the responsibility to receive impulses from other neuron's axons terminals. Dendrites are built of multiple branches that provide extended surface for the neuron to receive multiple impulses. This extended surface provides the opportunity for the neuron to receive many signals from many sources. Impulse which is transferred from one neuron's axon to the next neuron's dendrite has to pass through synapses. Due to the strength of the signal and also size of the gap between the axon's terminal and the next dendrite, travelling impulse either fades out, stays unchanged or become stronger. Signal then will travel through dendrite and enters the soma, excites it, and causes a signal through this neuron's axon to be transferred to the next neurons. The more impulses (representing proper path of action and reaction) travel through a dendrite through the time, the stronger the dendrites gets. The current general scientific community belief is that the memory of human develops and locates here. Understanding this system has been the main source for enthusiastic engineers and mathematicians to mimic its behavior in generating artificial neural network algorithm which is one of the most powerful mega-data machine learning tools in order to generate machines (artificial intelligent machines) that can learn from experience and change their behavior accordingly (Fig. 14.1).

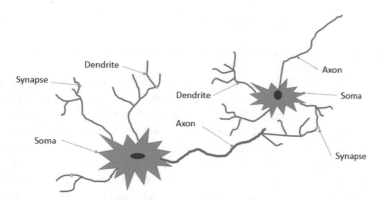

Fig. 14.1 Nervous system structure

An artificial neural network algorithm is generally made of at least two and in most of the cases one or two additional layers. The two layers far on each sides represent the input and output layers of the artificial neural network algorithm. The intermediate layers when they exist are called hidden layers. The whole model can be explained as similar to a set of nerve cells interconnecting to develop a nervous system. The input layer can be assumed to operate like dendrites, and is made of multiple starting calculating units receiving the inputs from one end and deliver their output to the next layer. A cell can receive multiple different signals, but it can only provide one output signal (similar to axon). This output stays unchanged and can be delivered to multiple downstream cells in the next layer with the same exact strength. The strength of the connecting line between two cells (neurons) can be modeled similar to synapsis which represents the strength of the connections between the neurons. Every time that nervous system generates a correct response to a given input it will strengthen the connections and every time the nervous system generates a wrong response to a given input it will weaken the connections. As the system receives repeating weight adjustment the connections become stronger or weaker and therefore the process which is called learning process continues.

In 1943, Warren McCulloch and Walter Pitts proposed a very simple idea that is still the basis for most artificial neural networks. The neuron computes the weighted sum of the input signals and compares the result with a threshold value, θ. If the net input is less than the threshold, the neuron output is -1. But if the net input is greater than or equal to the threshold, the neuron becomes activated and its output attains a value +1 [1, 2].

The above function is called sign function. Other similar useful functions are step, linear, sigmoid and tangent hyperbolic functions. Sigmoid functions has a great use in solving artificial neural network problems. It can accept any value between $+-\infty$ and delivers answers in the range of 0 and 1 and is defined as:

$$Y_{\text{Sigmoid}} = 1/(1 + e^{-x}) \tag{14.1}$$

Tangent hyperbolic also can accept any value between $+-\infty$ and delivers answers in the range of -1 and 1.

Let's see how the algorithm should be set up and how it works. Even though the number of inputs, hidden layer joints and outputs can be as many as the designer desire let's assume we have a set of three inputs, five hidden layer joints and two outputs. Input values are assigned to be $x1$, $x2$, and $x3$, hidden layer joint values are named $h1$, $h2$, $h3$, $h4$, and $h5$ and the output values are $o1$ and $o2$. The connection weight between input number 1 and hidden layer joint number 1 is called $wx1h1$, the connection weight between input number 2 and hidden layer joint number 1 is called $wx2h1$, and the connection weight between hidden layer number 1 and output layer joint number 1 is called $wh1o1$, etc. The bias or threshold value in each hidden layer and output joints are called $\theta1$, $\theta2$, etc. Also assume the selected function for each cell (joint) in hidden layer and output layer has been designed based on the Sigmoid function. Now let's restate the objective of our experiment here again. We have a set of existing data that each set is representing with the associated values for $x1, x2, x3, o1$, and $o2$. There is a long list of these related items and we are searching for a pattern (pattern recognition) in these data in order to be able to predict the future possible outputs if we know the new input conditions (Fig. 14.2).

We proceed first with calculating the results of functions for $h1$, $h2$, $h3$, $h4$, and $h5$ and then use the results to calculate the results of functions for $o1$ and $o2$. In order to be able to do these calculations, we need to make a random selection for the weights of connections and biases (thresholds) for hidden and output layer nodes. Random selection of the weights and biases causes a difference between the real outputs ($o1$ and $o2$) and the calculated outputs which is called the error in outputs. The most popular method for decreasing the size of the error down to an acceptable level is using back-propagation learning law. By using this method and based on the calculated error in outputs, we can recalculate the weights and biases in each joints.

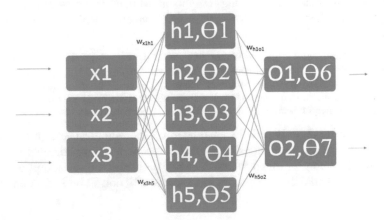

Fig. 14.2 General structure of artificial neural network algorithm

The next step is to use these new weights in calculating the second set of data, check the error and revise the weights with back-propagation method again. We will repeat this procedure through all the learning sets and polish the weights and biases each time. The process can be stated as follows:

First we should calculate processors at $h1$, $h2$, $h3$, $h4$, and $h5$ in a similar manner to Eq. 14.2 to 14.6 below:

$$h1 = 1/\left(1 + e^{-(x1 \times wx1h1 + x2 \times wx2h1 + x3 \times wx3h1 - \theta1)}\right) \tag{14.2}$$

$$h2 = 1/\left(1 + e^{-(x1 \times wx1h2 + x2 \times wx2h2 + x3 \times wx3h2 - \theta2)}\right) \tag{14.3}$$

$$h3 = 1/\left(1 + e^{-(x1 \times wx1h3 + x2 \times wx2h3 + x3 \times wx3h3 - \theta3)}\right) \tag{14.4}$$

$$h4 = 1/\left(1 + e^{-(x1 \times wx1h4 + x2 \times wx2h4 + x3 \times wx3h4 - \theta4)}\right) \tag{14.5}$$

$$h5 = 1/\left(1 + e^{-(x1 \times wx1h5 + x2 \times wx2h5 + x3 \times wx3h5 - \theta5)}\right) \tag{14.6}$$

The next step is to calculate $o1$ and $o2$ in a similar manner as follows:

$$o1 = 1/\left(1 + e^{-(h1 \times wh1o1 + h2 \times wh2o1 + h3 \times wh3o1 + h4 \times wh4o1 + h5 \times wh5o1 - \theta6)}\right) \tag{14.7}$$

$$o2 = 1/\left(1 + e^{-(h1 \times wh1o2 + h2 \times wh2o2 + h3 \times wh3o2 + h4 \times wh4o2 + h5 \times wh5o2 - \theta7)}\right) \tag{14.8}$$

Therefore the error for output number 1 and 2 can be calculated as follows:

$$e1 = \text{desired} \quad o1 - o1 \tag{14.9}$$

$$e2 = \text{desired} \quad o2 - o2 \tag{14.10}$$

At this point the process of weight correction (back-propagation) should be done to decrease the error size as follows:

$$\text{Error Gradient} \quad \text{for} \quad o1 = o1 - (1 - o1) \times e1 \tag{14.11}$$

$$\text{Error Gradient} \quad \text{for} \quad o2 = o2 - (1 - o2) \times e2 \tag{14.12}$$

And all weights and biases values at joint $o1$ and $o2$ will have corrections similar to what is depicted in Eqs. 14.13 and 14.14 based on a given learning rate (e.g., 0.1 or 1):

Correction for $wh1o1 = (\text{Rate of learning}) \times h1 \times (\text{ErrorGradient for } o1)$ (14.13)

Correction for $\theta 6 = (\text{Rate of learning}) \times \theta 6 \times (\text{Error gradient for } o1)$ (14.14)

And the corrections for weights and biases between hidden layer and input layer will be calculated similar to what is presented in Eqs. 14.15, 14.16, and 14.17:

$$\text{Error Gradient for } h1 = h1 - (1 - h1) \times (\text{Error Gradient for } o1) \\ \times wh1o1 \quad (14.15)$$

Correction for $wx1h1 = (\text{Rate of learning}) \times x1 \times (\text{ErrorGradient for } h1)$ (14.16)

Correction for $\theta 1 = (\text{Rate of learning}) \times \theta 1 \times (\text{Error gradient for } h1)$ (14.17)

The updated weights and biases should be used for next set of data and at the end of first run. The process then will be repeated until we reach the desired threshold. The final set of weights and biases then will be used for predicting new outputs. For a detail example of this approach see Negnevitsky [2]. For a detail representation of back-propagation technique and also a simple calculated example using back-propagation method see the links below:

http://page.mi.fu-berlin.de/rojas/neural/chapter/K7.pdf
http://www.dataminingmasters.com/uploads/studentProjects/NeuralNetworks.pdf

It can be seen when there is a large set of data for the process of learning the above method (back-propagation) may become very time consuming and impractical. To solve this problem different evolutionary methods have been presented. One of the most attractive method that can complement artificial neural network is utilizing a genetic algorithm in order to optimize the multipliers (weights and thresholds (biases)). For example we can use a simple tool such as Microsoft excel solver add-in to solve the artificial neural network problems similar to the approach that is taken for a single optimization problem. The main work then is to develop the weights and thresholds (genes) to form a chromosome and solve it via methods represented in previous chapters. The fitness of the chromosome here will be the average of square root of the sum of the square of the errors after each full run. The smaller the error the fitter the chromosome will be.

Similar to the other sections of the book let's continue with an application of this technique in an HVAC format. To do so assume we have a long list (here for the benefit of time we suffice on only 25 data set) of the existing data gathered from different buildings (the more data we can gather the higher the reliability of our calculations would be). The data gathered are $x1$ (quantity of outside air flow per person designed for the building (cubic foot/min), $x2$ (the average window to wall ratio of the building (sq ft/sq ft), $x3$ (the average ceiling height of the building (ft)) and $o1$ (the yearly energy consumption per square foot of the building (1000 btu/year/sq ft)). It has to be noted that the data used here are estimates and not from a

	Existing Data	Existing Data	Existing Data		Existing Data
	x1	x2	x3		Output
	5	0.4	10		25.1403103
	6	0.43	11		34.65382221
	7	0.6	12		45.90476755
	6.5	0.7	12		40.34309802
	5.5	0.55	13		30.57272385
	7.5	0.54	9		50.91144638
	8	0.33	8		56.9238616
	7.25	0.67	9.5		48.08743699
	6.5	0.7	11		40.0158387
	8	0.6	10.5		57.91166551
	5.2	0.5	10		26.88836652
	6	0.55	9		34.04199652
	5	0.7	10		25.30756558
	6	0.6	9		34.07088358
	5	0.5	10		25.18836652
	7	0.55	11		45.54862118
	8	0.6	9		57.40421691
	7.5	0.5	10.5		51.39746358
	6.5	0.45	9		39.19831832
	5	0.6	10.5		25.41166551
	6	0.55	9		34.04199652
	5	0.45	10		25.16333653
	5.25	0.55	9.5		27.18165085
	6.75	0.6	9		42.03963358
	7.5	0.65	10		51.31665329
min	5	0.33	8		25.1403103
max	8	0.7	13		57.91166551

Chart 14.1 Sample of existing data set (three inputs and one output)

real source and also the higher the factors we choose as input data (e.g., $x4$, $x5$, ...)
our results would be more reliable.

These data set would be used to educate us about the relationship between these
inputs and therefore helps us to predict the possible future occasions. Charts 14.1
and 14.2 show the listed existing data and also the same data in a normalized format
in order to make all the input and output to fit within 0 and 1 margin.

Normalized X1	Normalized X2	Normalized X3		Normalized Output
0	0.18918919	0.4		0
0.33333333	0.27027027	0.6		0.29029962
0.66666667	0.72972973	0.8		0.63361607
0.5	1	0.8		0.46390476
0.16666667	0.59459459	1		0.16576713
0.83333333	0.56756757	0.2		0.78639214
1	0	0		0.9698577
0.75	0.91891892	0.3		0.70021903
0.5	1	0.6		0.45391862
1	0.72972973	0.5		1
0.06666667	0.45945946	0.4		0.05334098
0.33333333	0.59459459	0.2		0.27163009
0	1	0.4		0.0051037
0.33333333	0.72972973	0.2		0.27251156
0	0.45945946	0.4		0.00146641
0.66666667	0.59459459	0.6		0.62274846
1	0.72972973	0.2		0.98451548
0.83333333	0.45945946	0.5		0.80122269
0.5	0.32432432	0.2		0.42897243
0	0.72972973	0.5		0.00828026
0.33333333	0.59459459	0.2		0.27163009
0	0.32432432	0.4		0.00070263
0.08333333	0.59459459	0.3		0.06229039
0.58333333	0.72972973	0.2		0.51567362
0.83333333	0.86486486	0.4		0.79875681

Chart 14.2 Sample of the existing data set (normalized)

The next step is to set the hidden and output layers joints to execute the Sigmoid function while using the assigned weight and bias values, and check the calculated errors versus the existing associated data set output. The assigned weights and biases should be randomly assumed at the beginning of the simulation.

These numbers will be tuned up during the optimization procedure gradually until the error size reaches the acceptable level. The final weights and biases at the end of the process are the weights and biases which should be used in estimating the future choices. That's how learning process would lead to prediction of the output value. See Charts 14.3, 14.4, and 14.5 for the process of calculating the weights and

Based on Sigmoid Function			Based on Sigmoid Function			
h1	h2	h3	Output	DIFF	ERROR	RMS
0.781358	5.83E-05	0.993052	0.013726009	-0.01373	0.000188403	0.014351
0.107836	0.001101	0.959381	0.298546214	-0.00825	6.80063E-05	
0.003341	0.018975	0.789867	0.626580315	0.007036	4.95018E-05	
0.01385	0.004019	0.895312	0.477668026	-0.01376	0.000189427	
0.249885	0.000196	0.977903	0.163609976	0.002157	4.65332E-06	
0.001356	0.101864	0.684493	0.800667326	-0.01428	0.000203781	
0.00046	0.382885	0.519087	0.968721713	0.001136	1.29047E-06	
0.002245	0.045672	0.753402	0.692668304	0.007551	5.70135E-05	
0.016785	0.004327	0.903835	0.464639677	-0.01072	0.000114941	
0.000193	0.308932	0.442223	0.969397628	0.030602	0.000936505	
0.622936	0.000101	0.990008	0.029494602	0.023846	0.00056865	
0.130933	0.001198	0.965062	0.27053974	0.00109	1.18887E-06	
0.700192	4.98E-05	0.992474	0.020249487	-0.01515	0.000229395	
0.123074	0.001167	0.964606	0.278506531	-0.00599	3.59396E-05	
0.756171	5.53E-05	0.992864	0.015490967	-0.01402	0.000196688	
0.004355	0.02094	0.807205	0.608557862	0.014191	0.000201373	
0.000258	0.333195	0.477378	0.967775883	0.01674	0.000280214	
0.001071	0.093896	0.655579	0.816034568	-0.01481	0.000219392	
0.034711	0.005719	0.923899	0.421877857	0.007095	5.0333E-05	
0.709383	5.06E-05	0.99232	0.019399074	-0.01112	0.000123628	
0.130933	0.001198	0.965062	0.27053974	0.00109	1.18887E-06	
0.769005	5.68E-05	0.992959	0.014565474	-0.01386	0.000192178	
0.591774	0.000119	0.989489	0.034214014	0.028076	0.000788283	
0.013057	0.011206	0.884093	0.499735898	0.015938	0.000254011	
0.000955	0.090392	0.657134	0.812631952	-0.01388	0.00019252	

Chart 14.3 Calculated hidden and output layer values and output error size

biases and predicting the energy consumption per year per square foot of a given building by knowing the outside airflow capacity, ratio of the window to wall, and ceiling height of the under question building. This is one of the most important methods that currently is used for machine learning and robotics.

Chart 14.4 Weight and
bias chromosome

wi1h1	-9.44759
wi2h1	-0.5247
wi3h1	-0.97626
wi1h2	9.088032
wi2h2	-0.19491
wi3h2	-0.3712
wi1h3	-5.09411
wi2h3	-0.09942
wi3h3	-0.47203
wh1o1	-4.84033
wh2o1	4.52725
wh3o1	-4.63228
Teta 1	1.763499
Teta 2	-9.5654
Teta 3	5.170507
Teta 4	4.106775

xi	wxihj	hi	whioj	Oi
		1.763499		4.106775
0.55	-9.44759	0.019171		
	9.088032		-4.84033	
	-5.09411			
		-9.5654		
0.4	-0.5247	0.008528	4.52725	0.471974
	-0.19491			
	-0.09942			
		5.170507		
0.3	-0.97626	0.899089		
	-0.3712		-4.63228	
	-0.47203			

Chart 14.5 Calculated weights and biases used to predict new set output

References

1. McCulloch, W. S., & Pitts, W. (1943). A logical calculus of the ideas immanent innervous activity. *Bulletin of Mathematical Biophysics, 5*, 115–137.
2. Negnevitsky, M. (2005). *Artificial intelligence, a guide to intelligent systems*. Boston, MA: Addison-Wesley.

Chapter 15
Fuzzy Logic

Abstract Even though the concept of fuzzy logic has been developed in the 1920s, but it was Lotfi A. Zadeh in University of California, Berkeley in 1965 that proposed the fuzzy set theory for the first time. The traditional Boolean logic offers only either 0 or 1 (false or true) as the acceptable values for a given variable. To the contrary fuzzy logic can offer all the possibilities between 0 and 1 as the assigned truth value of the variables.

Keywords Fuzzy logic • Boolean logic • Fuzzy representation • Comfort level • Humidity ratio • Temperature • Fuzzification • Defuzzification • Discomfort level • Metabolism

Even though the concept of fuzzy logic has been developed in 1920s, but it was Lotfi A. Zadeh in University of California, Berkeley in 1965 that proposed the fuzzy set theory for the first time. The traditional Boolean logic offers only either 0 or 1 (false or true) as the acceptable values for a given variable. To the contrary fuzzy logic can offer all the possibilities between 0 and 1 as the assigned truth value of the variables. These values in fact represent a degree of membership of the variable to a given category. In traditional Boolean logic we usually draw a straight dividing line and sharply divide the characteristics of the events or objects such as when we set a crisp line of 200 pounds to separate overweight and not overweight persons for a specific category, it then implies that if somebody in this category is even 199 pounds he would not be considered as overweight, while a person who is 201 pound is considered to be overweight. Fuzzy logic helps us to assign degree of membership to the overweightness and not overweightness of people. Therefore, for example we can say a person in a specific category who is 190 pounds has a 25 % membership to the overweigh category and 75 % membership to not overweight category and instead a person who is 210 pounds has a 90 % membership to the overweight category and 10 % membership to the not overweight category.

Furthermore and despite of the fact that both probability (in statistics) and truth value (in fuzzy logic) have ranges between 0 and 1, but the former represents the degree of lack of knowledge in our mathematical models (see uncertainty chapter in this book), while the latter represents the degree of vagueness of the phenomenon depicted by the mathematical model. As an example, the fact that prevents us from predicting exactly there will be snow in next few hours is due to our lack of knowledge (uncertainty: there is 50 % chance that it snow in next few hours), but the reason that we cannot surely state it is currently snowing is the fact that essentially we cannot distinguish between ice rain and snow with certainty (fuzzy logic: what is falling down from the sky now is 40 % ice rain and 60 % snow). It therefore implies that even though the uncertainty and fuzzy state both are represented by values between 0 and 1 there is still a major difference between these two concepts. As another example, when we are talking about the existence of a building and we think about if the owner's loan for construction will be approved, if the state of the economy will justify construction of this specific type of building in a near future or if owner's desire to construct this type of building will stay strong to go through the whole construction process we are dealing with uncertainty and as we discussed in decision-making under uncertainty we can calculate and assign the degree of possibility for this building to be built in the near future or not. On the other hand, when we are talking about the existence of the building in fuzzy world, we could be talking about at what point during the construction we can call the structure a building. Is it fine to call the structure when the concrete slab is poured a building, or the structure should have at least the surrounding walls and roof before we call it a building? Is it justified to call it a building when the plumbing and air conditioning are installed or we have to wait until it is completely ready to be occupied before call it a building? What is obvious that at different points from the time that the site is becoming ready for the building to be constructed until the time that building is completely ready to be occupied we can have different perceptions of this structure and levels of membership of it as being a building. If we assign zero membership to the site and 100 % membership to the fully ready to be occupied building we can see at each step in between we may have a partial level of membership for this structure to be called a building, e.g., 0.3 membership to being a building when the walls and roof are installed, and 0.7 membership when the plumbing and air conditioning are installed. Such membership assignment to different levels of a phenomena or object is in the heart of fuzzy logic concept. Fuzzy logic has multiple applications in different fields of science and engineering including artificial intelligence and controls. Some of successful implementations of fuzzy logic have been preventing overshoot–undershoot temperature oscillation and consuming less on–off power in air conditioning equipment, mixing chemicals based on plant conditions in chemical mixers, scheduling tasks and assembly line strategies in factory control and adjusting moisture content to room conditions in humidifiers. For an extended list of successful applications of fuzzy logic rules see http://www.hindawi.com/journals/afs/2013/581879/ and http://www.ifi.uzh.ch/ailab/teaching/formalmethods2013/fuzzylogicscript.pdf.

As another example let us assume we are defining the air temperature in a space with a Boolean logic as higher than 78 °F as hot air and lower than 64 °F as cold air. Therefore, the medium temperature region will fall in between 64 and 78 °F. See Fig. 15.1 below.

Now let us switch our approach from Boolean to fuzzy logic method and reformat the temperature regions in the space to fit the premises of this method.

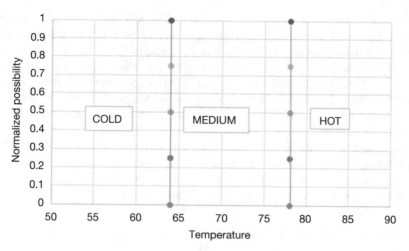

Fig. 15.1   Boolean representation of cold, medium, and hot temperatures

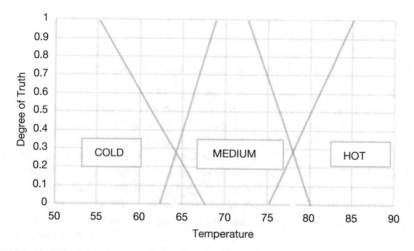

Fig. 15.2   Fuzzy representation of cold, medium, and hot temperatures

Therefore, with fuzzy logic we have divided the space into three neighborhoods of cold, medium, and hot temperature zones as something similar to what is shown in Fig. 15.2 above. It can be noticed that we have created some areas of overlap between cold and medium and also hot and medium areas. In such representation if we draw a vertical line from 60 °F and extend it towards the top of the chart it will intersect the line representing margin of the cold temperature at approximately 60 % of the degree of the truth, and if we draw another vertical line from temperature 78 °F and extend it towards the top of the chart it will intersect the margins of medium and hot air areas at approximately 10 % and 40 % respectively. These values represent the degree of membership of 60 and 78 °F to cold region, medium region and hot region as 60 %, 10 % and 40 % respectively. Other readings from Fig. 15.2 can be defined as 55 °F and below have a 100 % membership for coldness, 68 °F has a 0 % membership for cold, 63 °F has a 0 % membership for medium, 68–73 °F have 100 % membership for medium temperature, 80 °F has a 0 % membership for medium, 75 °F has a 0 % membership for hot, and 83 and higher have 100 % membership for hotness.

As it was noted earlier, in fuzzy theory a fuzzy set F can be defined by a membership function mF(x), where mF(x) is equal to 1 if x is completely in F, is 0 if x is not part of F, and is between 0 and 1 if x is partly in F.

So if we go back to two figures represented above and if in Boolean logic we define three subsets of:

B1 Cold = {$x1, x2, x3$} = {50, 55, 60}
B2 Medium = {$x1, x2, x3$} = {65, 70, 75}
B3 Hot = {$x1, x2, x3$} = {80, 85, 90}

Then we can define the same subsets in a fuzzy logic context as:

F1 Cold = {($x1$, mF($x1$)), ($x2$, mF($x2$)),($x3$, mF($x3$))} = {(50, 1), (55, 1), (60, 0.6)}
F2 Medium = {($x1$, mF($x1$)), ($x2$, mF($x2$)),($x3$, mF($x3$))} = {(65, 0.4), (70, 1), (75, 0.7)}
F3 Hot = {($x1$, mF($x1$)), ($x2$, mF($x2$)),($x3$, mF($x3$))} = {(80, 0.5), (85, 1), (90, 1)}

In fuzzy logic, linguistic variables act as fuzzy variables. When we instead of saying if the temperature is above 80 °F the weather is hot, say if the temperature is high the weather is hot we are using fuzzy language to define the temperature of the weather. If we understand that the possible weather temperatures occurring in a region is between 10 and 100 °F in order for a given variable in this range to be represented by fuzzy linguistic wording, we use hedges such as very and extremely.

In fuzzy logic there are rules to change the statement containing linguistic hedges to mathematical equivalents of non-hedge expressions, such as when we instead of high membership mF($x1$) say very high membership we can use $[mF(x1)]^2$, and when we instead of low membership mF($x2$) say extremely low membership we can use $[mF(x2)]^3$. "When we define the fuzzy sets of linguistic variables, the goal is not to exhaustively define the linguistic variables. Instead, we only define a few fuzzy subsets that will be useful later in definition of the rules that we apply it" [1]. For a detail set of these conversion hedges refer to several available references such as [2] which expresses the fuzzy logic procedure in detail.

Similar to classical set theory, there are some operations (interaction) functions that are typically used in fuzzy logic theory. The most used operations in fuzzy logic are AND, OR, and NOT. NOT is defined as "NOT mF($x1$) = 1 − mF($x1$)," while AND is defined as selecting the maximum value between two variables and OR is selecting the minimum value between two variables. In many cases we may confer with situations that the antecedent (IF) or consequent (Then) portions of the rule are made of multiple parts such as:

IF Temperature is Medium
AND Humidity is low
AND Number of people in room is very low
Then Working condition is comfortable
AND Number of complaints is low

In fuzzy logic, the process of mapping from input(s) to output(s) is called fuzzy inference. Different methods of fuzzy inference have been developed. One of the most popular methods of fuzzy inference is Mamdani style. In this style the designer at the beginning fuzzifies the clear input variables, then evaluates the rules, aggregates the rules and finally defuzzificates the results to reach a clear answer. Jager [3] criticizes the method due to its similarity to a general interpolation during defuzzification. "Because of the defuzzification, a fuzzy controller (or more general, a fuzzy system) performs interpolation between tuples in a hyperspace, where each tuple is represented by a fuzzy rule. Therefore, a fuzzy controller can be "simplified" to a combination of a look-up table and an interpolation method" [3]

To put the steps in perspective, let us assume we are making a fuzzy logic problem including three inputs and one output.

Hypothetically let us assume we are planning to define a control algorithm for measuring the discomfort level of the occupants in a space based on the room temperature, room humidity and level of activity of the occupants. This will be based on input form temperature, humidity and motion sensors. Then instead of operating the air handling unit system that is serving this space solely based on input from the temperature (and in some case humidity) sensor(s), operate it based on using the aggregated discomfort level developed by all these three inputs. Fuzzy logic method can be a very good approach for developing such control system. The first step is to define the fuzzy spaces that are associated with our linguistic definitions such as low, medium and high for the input and output variables. In order to make the understanding of the system easier let us limit our variables just to the three main variables expressions and do not include the hedged variables such as very and extremely in our work. Assume our expert in the HVAC domain has laid out the fuzzy spaces for our three input variables (temperature, humidity ration, and metabolism) and the output variable (discomfort level) similar to Figs. 15.3, 15.4, 15.5, and 15.6. It can be seen that each variable is divided to regions of low, medium, and high with linear, triangular, or trapezoidal areas. "The shape of the membership function is chosen arbitrarily by following the advice of the expert or by statistical studies: sigmoid, hyperbolic, tangent, exponential, Gaussian or any other form can be used" [1]. The overlaps between each two spaces are where the variable can be a member of either side with different degree of membership.

Let us assign three rules for this control system to be followed: (It should be noted that we can have as much as rules that we think is required to express the field accurately.)

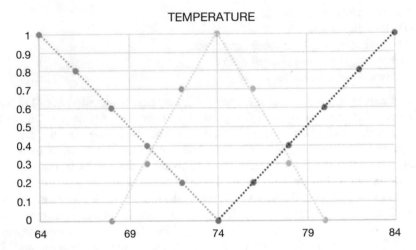

Fig. 15.3 Fuzzy presentation of temperature

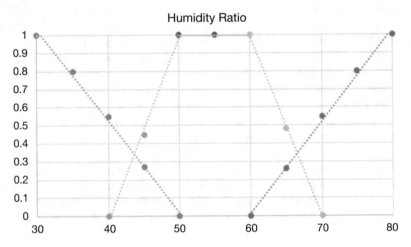

Fig. 15.4 Fuzzy presentation of relative humidity

Rule 1
IF "Temperature is Low AND Humidity Ratio is Low" AND Metabolism is High
Then Discomfort level is Low
IF "Temperature is Low AND Humidity Ratio is High" OR Metabolism is High
Then Discomfort level is Medium
IF "Temperature is Low AND Humidity Ratio is Low" OR Metabolism is Low
Then Discomfort level is High

Now assume the non-fuzzy input variables that we have to work with are 67 °F,
34 % relative humidity, and 1.29 met. It can be seen that either of these inputs if

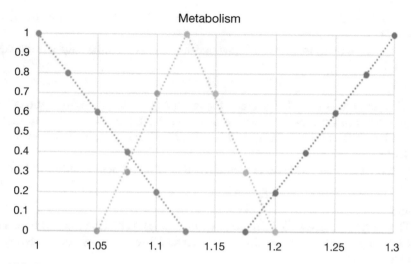

Fig. 15.5 Fuzzy presentation of metabolism

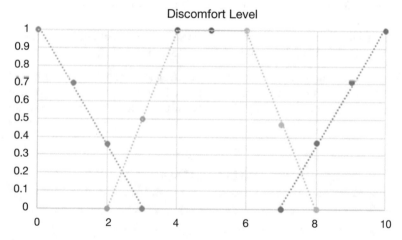

Fig. 15.6 Fuzzy presentation of discomfort level

extended vertically in its associated chart would intersect low, low, and high fuzzy separation lines in Figs. 15.3, 15.4, and 15.5 above accordingly. The associated membership values for these inputs are 0.7 for temperature, 0.8 for humidity ratio, and 0.9 for metabolism. The next step is to calculate the output fuzzy value based on the three rules that we have specified for our control system.

Based on rule 1;

Rule 1
IF "Temperature is Low (0.7 degree of membership) AND Humidity Ratio is Low (0.8 degree of membership)" AND Metabolism is High (0.9 degree of membership)

Which is equivalent to minimum of 0.7 and 0.8 (0.7) and then minimum of 0.7 and 0.9 (0.7).

Then correct Discomfort level is Low and its level of membership will be equal to 0.7.

Rule 2

IF "Temperature is Low (0.7 degree of membership) AND Humidity Ratio is High (0 degree of membership)" OR Metabolism is High (0.9 degree of membership)

Which is equivalent to minimum of 0.7 and 0.0 (0.0) and then maximum of 0.0 and 0.9 (0.9).

Then correct Discomfort level is Medium and its level of membership will be equal to 0.9.

Rule 3

IF "Temperature is Low (0.7 degree of membership) AND Humidity Ratio is Low (0.8 degree of membership)" OR Metabolism is Low (0.0 degree of membership)

Which is equivalent to minimum of 0.7 and 0.8 (0.7) and then maximum of 0.7 of 0.0 (0.7).

Then correct Discomfort level is High and its level of membership will be equal to 0.7.

The centroid of these three areas will represent the location of output variable which can be calculated and is equal to 5 (Figs. 15.7, 15.8, 15.9, and 15.10).

Therefore, based on our given inputs the level of discomfort would be medium and we can run the air handling unit to deal with this level of discomfort. As the inputs changes the same procedure shall be followed to calculate the new level of

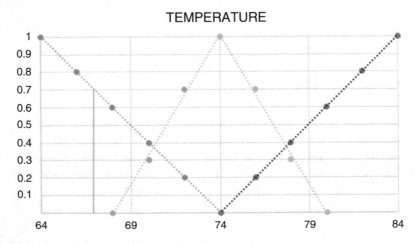

Fig. 15.7 Fuzzy presentation of 0.7 membership to low temperature

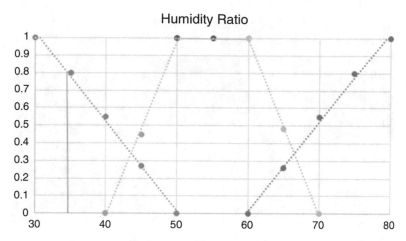

Fig. 15.8 Fuzzy presentation of 0.8 membership to low relative humidity

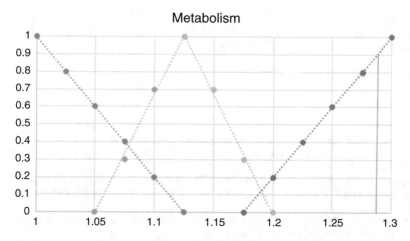

Fig. 15.9 Fuzzy presentation of 0.9 membership to high metabolism

discomfort and accordingly operate the air handling unit to control the discomfort level based on collection of these three inputs at each instant.

Mendel [4] notes that even though fuzzy logic and feed-forward neural network can both be used to solve similar problems, the advantage of fuzzy logic is its invaluable linguistic capability specifically when there is not a lot of numerical training data available.

> Fuzzy Logic provides a different way to approach a control or classification problem. This method focuses on what the system should do rather than trying to model how it works. One can concentrate on solving the problem rather than trying to model the system mathematically, if that is even possible [5].

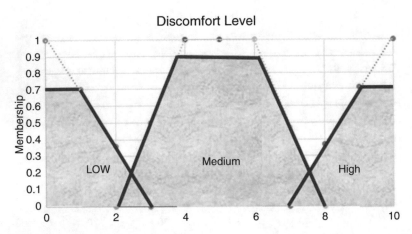

Fig. 15.10 Fuzzy presentation of overall discomfort level for selected temperature, relative humidity and metabolism

For a PowerPoint representation of the origins, concepts, and trends in fuzzy logic developed by professor Zadeh himself, see the following link http://wi-consortium.org/wicweb/pdf/Zadeh.pdf.

References

1. Dernoncourt, F. (2013). *Introduction to fuzzy logic*. Cambridge: MIT.
2. Negnevitsky, M. (2005). *Artificial intelligence, a guide to intelligent systems* (2nd ed.). Harlow: Addison-Wesley.
3. Jager, R. (1995). *Fuzzy logic in control*. PhD Thesis. Delft University of Technology, Delft.
4. Mendel, J. M. (1995). Fuzzy logic systems for engineering: a tutorial. *IEEE Proceedings, 83*, 345–377.
5. Hellmann, M. (2001). Fuzzy logic introduction. http://www.ece.uic.edu/~cpress/ref/2001-Hellmann%20fuzzyLogic%20Introduction.pdf

Chapter 16
Game Theory

Abstract Game theory is known to focus on the studying, conceptualizing, and formulating strategic scenarios (games) of cooperatively or conflicting decisions of interdependent agents. Describing the game requires mapping the players that are involved in the game, their preferences and also their available information and strategic actions and the payoff of their decisions.

Keywords Game theory • Cooperatively • Conflicting • Strategic choice • Nash bargaining • Coalitional game • Non-cooperative • Prisoner's dilemma • Head or tail game • Nash equilibrium • Player • Percent of mean vote • Percent of people dissatisfied

Game theory is known to focus on the studying, conceptualizing, and formulating strategic scenarios (games) of cooperatively or conflicting decisions of interdependent agents. Describing the game requires mapping the players that are involved in the game, their preferences and also their available information and strategic actions and the payoff of their decisions. "As a mathematical tool for the decision-maker the strength of game theory is the methodology it provides for structuring and analyzing problems of strategic choice. The process of formally modeling a situation as a game requires the decision-maker to enumerate explicitly the players and their strategic options, and to consider their preferences and reactions" [1].

The main difference between game theory and other methods that were represented under decision-making criteria in previous chapters of this book is that in other decision-making methods, the decision maker is only concerned about how he or she should make his or her decisions, but in the game theory the decision maker should be concerned about different agents that their decisions is going to affect his or her decision as well.

"Cooperative game theory encompasses two parts: Nash bargaining and coalitional game. Nash bargaining deals with situations in which a number of players need to agree on the terms under which they cooperate while coalitional game theory deals with the formation of cooperative groups or coalitions. In essence, cooperative game theory in both of its branches provides tools that allow the players to decide on whom to cooperate with and under which terms given several cooperation incentives and fairness rules" [2]. Since the best applications for the cooperative game theory usually are structured around the political or international relations type of sciences our general focus here will be only on the non-cooperative type games and therefore selection of strategic choices among the competitors or competing ideas.

Non-cooperative games in most literatures are divided to two sub-categories of static and dynamic games. In a static game each player acts once either simultaneously or in different times without knowing what the other player (players) does (do). In a dynamic game, players have some information about the choice of the other players in previous time and also will act multiple times. In some other literatures the non-cooperative games have been divided to four sub-categories: Normal, extensive, Bayesian, and repeated games. In a normal game the players act simultaneously without knowing the action of the other player, while in an extensive games players act in different orderly times. The Bayesian games are those that before the act, one player receives some information about the possible function of the other player, and repeated games are those that are repeated multiple times and therefore open the possibility of mapping the other player's moves. Either way, the fact is that under any of these definitions that we choose, the goal of each player is to try to choose his optimum possible action which is depended on both his and his opponent(s)' choices. The possible strategies that a player can have could be either pure (deterministic) or mixed (probability distribution). Nash equilibrium in non-cooperative games is defined as a state of equilibrium that none of the players unilaterally can improve their utility from the game by changing their strategy, if the other player's strategies remain unchanged. Number of the strategies should be finite and Nash equilibrium is only guarantied where mixed strategies are considered.

In other words in a general pure format a simple game between two agents can be described by the utilities that they receive based on their strategies. Assume Agents A & B each has two strategies that give them the utilities such as A1-B1 (6, 4), A1-B2 (5, 5), A2-B1 (10, 2), and A2-B2 (1, 1). If we evaluate three types of games, A move first, B move first, or neither knows about the other agents move and therefore move simultaneously, we will have the following possibilities. If A moves first, it is obvious that he is better off with his second strategy (utility of 10) and that forces B to choose his first strategy (utility of 2). If B moves first, it is obvious that he is better off with his second strategy (utility of 5) and that forces A to choose his first strategy (utility of 5). Both of these conditions are qualified as simple Nash equilibriums. On the other hand if both players are required to decide simultaneously, then we will have a mix Nash equilibrium. In this condition utilities that A and B will have will be 5.5 and 3 which comes from each player randomly plays his strategies with a 50–50 % chance.

Similarly if we assume Agents A & B each has two strategies that give them the utilities such as A1-B1 (6, 3), A1-B2 (5, 4), A2-B1 (9, 2), and A2-B2 (3, 1), then if A moves first it is obvious that he is better off with his second strategy (utility of 9) and that forces B to choose his first strategy (utility of 2). If B moves first, it is obvious that he is better off with his second strategy (utility of 4) and that forces A to choose his first strategy (utility of 5). Both conditions are again qualified as Nash

Table 16.1 Prisoner's dilemma game

Player 1 ↓/Player 2 ⟶	Cooperate	Defect
Cooperate	2, 2	0, 3
Defect	3, 0	1, 1

Table 16.2 Head or tail game

Player 1 ↓/Player 2 ⟶	Head	Tail
Head	1	−1
Tail	−1	1

equilibrium as well. And if both players are to decide simultaneously then we will have a mix Nash equilibrium again. In this condition utilities that A and B will have will be 5.1 and 2.5 which comes from player one randomly plays his strategies with a 50–50 % chance and player two plays his strategies with a 40–60 % chance. We will discuss such examples a little more later in this chapter.

Schecter and Gintis [3] define three uses for game theory: "(1) Understand the world. For example, game theory helps understand why animals sometimes fight over territory and sometimes don't. (2) Respond to the world. For example, game theory has been used to develop strategies to win money at poker. (3) Change the world. Often the world is the way it is because people are responding to the rules of a game. Changing the game can change how they act. For example, rules on using energy can be designed to encourage conservation and innovation."

Probably one of the most discussed examples of the game theory is the prisoner's dilemma. In this game two prisoners that are expected to have participated in the same crime are kept in separate rooms so they would not know what strategy the other person would choose while he is being questioned. Each prisoner has two possible strategies, either to cooperate with the other prisoner and not to confess to the crime, or defect the other prisoner and confess to the crime. Therefore the utility that the prisoners receive would be as such: (1) both cooperate and do not confess and each get a utility of equal to 2 or getting out of jail after a short time, (2) both defect and confess and each get a utility of equal to 1 or getting out of jail after a relatively longer time served, (3) one cooperate and the other one defect and therefore the one who has defected get to be freed immediately (utility of 3) and the one who has cooperated and has not confessed get the worst punishment to go to jail for a long time (utility of 0) (Table 16.1).

The dominant strategy here for both players is to defect, because for example the utility that player one gets by switching from defecting to cooperating while the other player stays with his original defect strategy, decreases from 1 to 0. Similarly the utility that player two gets by switching from his dominated strategy of defecting to cooperating while the first player stays with his original defect policy, decreases from 1 to 0 as well. This dominated strategy of (Defect, Defect) is also a saddle point (intersection between two defect strategies) and a Nash equilibrium. It should be noted here that strategy (Cooperate, Cooperate) in which the payoff of both players are higher than strategy (Defect, Defect) is only a Pareto optimal and is not to be considered as another Nash equilibrium state, since each player will be better off if switches his strategy assuming the other player does not change his.

Of course this is a static game and happens only once. But if it becomes a repeating game, and players can see what the other player did in previous games, they could change their strategy based on the history of the other player's strategies.

Now let us look at another well-known game (sum zero) where players 1 and 2 each has two options available based on Table 16.2.

In a sum zero game the amount of winning of one player is equal to the amount of loosing of the other player, therefore we can simplify the normal representation of the game as it is represented above. The game can be stated as two players will simultaneously show two coins and if both show the same side of the coin (head, head) or (tail, tail) the player one will win both coins, and if one shows head or tail and the other one shows the other side (tail or head) player number two will win both the coins. As it can be seen from the chart above there is no saddle point and no dominant strategy for either player that can be represented as Nash equilibrium condition, yet based on Nash equilibrium we should have at least one (pure or mix) equilibrium point. To find this state let us multiply first row by "p" and second row by "1−p" and first column by "q" and column two by "1−q" we can find a situation that would represent the state of Nash equilibrium. Then if player two keep playing strategy one (showing head all the time) or keep playing strategy two (showing tail all the time), the player one would have:

$$p\,(1) + (1 − p)(−1) = p\,(−1) + (1 − p)(1) \tag{16.1}$$

Which would lead to $p = 0.5$ and therefore $(1 − p) = 0.5$ as well. And if player one continuously play strategy one (showing head all the time) or keep playing strategy two (showing tail all the time), the player two would have:

$$q\,(1) + (1 − q)(−1) = q\,(−1) + (1 − q)(1) \tag{16.2}$$

Which would lead to $q = 0.5$ and therefore $(1 − q) = 0.5$ as well. This means the Nash equilibrium for both players is switching their strategy on a 50 % random selection. And at this state the payoff of each player would be zero.

In a good attempt for utilizing game theory method in HVAC and energy conservation Forouzandehmehr et al. [4] presented a solution for controlling HVAC systems in smart buildings using concepts of game theory. Their proposed learning algorithm makes it possible to minimize the cost of keeping maximum zones in the proper temperature range.

Coogan et al. [5] provided a framework for building energy management setting to incentivize desirable control policies among the building occupants based on

Table 16.3 Power company vs. customer game

PC ↓/CU ⟶	s1	s2	s3
S1	6	8	2
S2	−4	−6	6

Nash equilibrium and showed the technique on an eight zone building model can help to save around 3–4 % in energy cost.

At this point and similar to the other chapters in this book let us continue our discussion with an HVAC/ energy related problem and attempt to use the game theory in order to solve that. Now hypothetically assume a power company (PC) in regards to his customer (CU) switches between two strategies S1 and S2, and the customer has the three s1, s2, and s3 strategies when using the power supplied to its facility under assumption of sum zero evaluation. Assume the customer has a small onsite renewable energy production facility and an onsite energy storage as well. So at any instant he is capable of producing 10 % of the maximum facility power consumption. His strategies with a mixture of consumption control and utilizing renewable generation onsite are (s1) to keep the purchased consumption below X kWh, (s2) to keep the purchased consumption between X and Y kWh, and (s3) to allow the purchased consumption to be above Y kWh. On the other hand, the power provider has two strategies to apply (S1) a constant price \$C/kWh or (S2) a mix price structure to charge much lower when customer consumes above Y, moderate charge when customer consumes between X and Y, and very high charge when customer spend below X. Therefore, assume after considering the proper economic factors, such an assumption can lead to developing a decision table with sum zero utility such as what has been shown above (Table 16.3).

In order to find the equilibrium condition let us first find out if there is a dominant strategy for these players. If power provider (PC) selects strategy one (S1), he will be better off if customer (CU) selects either of strategies one or two (s1 or s2), but not better off if the customer (CU) selects strategy three (s3). Therefore the power company does not have a clear dominant strategy. Also Customer has no clear dominant strategy since (S1, s1) is better than (S1, s2) while (S2, s1) is worse than (S2,s2); (S1, s1) is worse than (S1, s3), while (S2, s1) is better than (S2, s3); and (S1, s2) is worse than (S1, s3), while (S2, s2) is better than (S2, s3). Therefore there is no pure Nash equilibrium in this problem. Also we can reach the same conclusion if we calculate the Maxi-min or maximum of the minimums "2" of rows and Mini-max or minimum of the maximums of columns "6". In sum zero games while the strategy of the first player is to maximize the minimum of his gain, the strategy of the second player is to minimize the maximum possible gain of the first player which would be his loss as well.

That forces us in order to find the Nash equilibrium use a mix strategy. To do this let us assume the probabilities of selecting S1 and S2 by the power company are $p1$ and $p2$ (while $p1 + p2 = 1$) and the probability of selecting s1, s2, and s3 by the customer are $q1$, $q2$, and $q3$ (while $q1 + q2 + q3 = 1$). If we try to minimize the maximums of the columns and maximize the minimums of the rows ($6p1-4p2$; $8p1-6p2$; $2p1 + 6p2$; $6q1 + 8q2 + 2q3$; $-4q1 - 6q2 + 6q3$) we will find out the conditions for Nash equilibrium in this situation is met where $p1 = 5/7, p2 = 2/7$, $q1 = 5/7$, $q2 = 0$ and $q3 = 2/7$. Therefore the equilibrium solution for this problem would be a plan that power company uses 5/7 of times strategy S1 and 2/7 of times strategy S2, while the customer uses his strategy s1 (keep the purchased consumption below X kWh) for 5/7 of the times and strategy s3 (allow the

Table 16.4 Temperature–
relative humidity 2 × 2
Strategy—Player 1

PPD/PMV (°F)	45 % RH	50 % RH
74	0.22/0.001	0.19/0.06
75	0.07/0.05	0.03/0.05

Table 16.5 Temperature–
relative humidity 2 × 2
Strategy—Player 2

PPD/PMV (°F)	45 % RH	50 % RH
74	0.04/0.05	0.01/0.05
75	0.1/0.05	0.13/0.05

Table 16.6 Temperature–
relative humidity 2 × 2
Strategy—Utilities

	Player 2	45 % RH	50 % RH
Player 1		$q1$	$1 - q1$
74 °F	$p1$	9.2, 6.2	4.5, 7.6
75 °F	$1 - p1$	5.6, 5.3	6.5, 5

purchased energy to be above Y kWh) for 2/7 of times, and do not use his strategy s2 at all.

Now let us discuss another condition in which we want to control the temperature and relative humidity in order to maximize comfort level in an office space for two people (player 1 and Player 2) that work in this office next to each other. Assume the first Player has a metabolism "met" equal to 1.05 and a "clo" equal to 0.8 while the second player has a "met" equal to 1.3 and a "clo" equal to 0.6. In order to simplify the process assume we are trying to control the air handling unit which is serving this space in order to choose strategies among 74 and 75 °F ambient (assume same as mean radiant temperature) and 45 and 50 % relative humidity combination to maximize the comfort level of both players. First if we use the ASHRAE developed PMV-PPD calculator to calculate each person percent of mean vote (PMV) and percent of dissatisfied values (PPD) in each instant we can generate two tables similar to the tables shown above (Table 16.4 Player 1 and Table 16.5 Player 2):

I have used the absolute value of PPD since in our experience it does not matter if PPD is positive or negative values. Now let us define a utility for level of comfort of each player by defining this utility as natural logarithm of reverse value of PPD multiplied by PMV. In this way the absolute value of multiplication of PPD and PMV creates a measurement for level of discomfort, by reversing this value we attempt to create a measurement for level of comfort and by taking the natural logarithm of that we attempt to create a utility value for this calculated comfort level. These utility values are depicted in Table 16.6 above.

To continue, first let us check to see if there is a pure Nash equilibrium condition available in this condition. With a little investigation we can find out for Player 1, if he switches from his first strategy 74°F he would be better off if the Player 2 is selecting his second strategy or 50 % RH, but worse off if Player 2 is selecting his first strategy or 45 % RH. Similarly if Player 2 switches from his first strategy 45 %

RH he would be better off if the Player 1 is selecting his first strategy or $74°$F, but worse off if Player 1 is selecting his second strategy or $75°$F.

Now that there is not a pure Nash equilibrium available, let us try to find a mix strategy by assigning $p1$ and $1 - p1$ probability distribution to Player's 1 strategies and $q1$ and $1 - q1$ probability distribution to Player's 2 strategies. Each Player is only willing to mix his strategies if the utility that he would get from either of his strategies are the same. Therefore, the following conditions should stand:

For Player 1; $P1 \times 9.2 + (1 - p1) \times 5.6$ (if the Player 2 chooses his first strategy) should be equal to $p1 \times 4.5 + (1 - p1) \times 6.5$ (if the Player 2 chooses his second strategy); and for Player 2; $q1 \times 6.2 + (1 - q1) \times 7.6$ (if the Player 1 chooses his first strategy) should be equal to $q1 \times 5.3 + (1 - q1) \times 5$ (if the Player 1 chooses his second strategy).

Table 16.7 Temperature–relative humidity 3×3 Strategy—Player 1

met	1.4		40 % RH		50 % RH		60 % RH	
clo	0.5		PMV	PPD	PMV	PPD	PMV	PPD
		73.5 °F	−0.13	0.05	−0.07	0.05	0.01	0.05
		74 °F	−0.06	0.05	−0.01	0.05	0.05	0.05
		74.5 °F	0.01	0.05	0.06	0.05	0.15	0.05

Table 16.8 Temperature–relative humidity 3×3 Strategy—Player 2

met	1.2		40 % RH		50 % RH		60 % RH	
clo	0.7		PMV	PPD	PMV	PPD	PMV	PPD
		73.5 °F	−0.16	0.06	−0.1	0.05	−0.03	0.05
		74 °F	−0.09	0.05	−0.03	0.05	0.04	0.05
		74.5 °F	−0.02	0.05	0.05	0.05	0.11	0.05

Table 16.9 Discomfort indicator 3×3 Strategy—Player 1

met	1.4		40 % RH	50 % RH	60 % RH
clo	0.5		DI = IPMVI × PPD	DI = IPMVI × PPD	DI = IPMVI × PPD
		73.5 °F	0.0065	0.0035	0.0005
		74 °F	0.003	0.0005	0.0025
		74.5 °F	0.0005	0.003	0.0075

Table 16.10 Discomfort indicator 3×3 Strategy—Player 2

met	1.2		40 % RH	50 % RH	60 % RH
clo	0.7		DI = IPMVI × PPD	DI = IPMVI × PPD	DI = IPMVI × PPD
		73.5 °F	0.0096	0.005	0.0015
		74 °F	0.0045	0.0015	0.002
		74.5 °F	0.001	0.0025	0.0055

Solving these equations leads to $p1$, $1 - p1$, $q1$ and $1 - q1$ to be equal to 0.16, 0.84, 0.59 and 0.41 respectively.

Therefore the best mix strategy would be to switch randomly 16 % and 84 % times between 74 and 75 °F and switch randomly 59 % and 41 % times between 45 % and 50 % RH. The utility that each Player gets from this strategy would then be:

Expected utility for Player $1 = (0.16 \times 0.59 \times 9.2) + (0.16 \times 0.41 \times 4.5) + (0.84 \times 0.59 \times 5.6) + (0.84 \times 0.41 \times 6.5) = 6.17$ and the expected utility

Table 16.11 Comfort utility 3 × 3 Strategy—Player 1

met	1.4		40 % RH	50 % RH	60 % RH
clo	0.5		LN(1/DI)	LN(1/DI)	LN(1/DI)
		73.5 °F	5.04	5.65	7.60
		74 °F	5.81	7.60	5.99
		74.5 °F	7.60	5.81	4.89

Table 16.12 Comfort utility 3 × 3 Strategy—Player 2

met	1.2		40 % RH	50 % RH	60 % RH
clo	0.7		LN(1/DI)	LN(1/DI)	LN(1/DI)
		73.5 °F	4.65	5.30	6.50
		74 °F	5.40	6.50	6.21
		74.5 °F	6.91	5.99	5.20

Table 16.13 Revised comfort utility 3 × 3 Strategy—Player 1

met	1.4		40 % RH	50 % RH	60 % RH
clo	0.5		Utility	Utility	Utility
		73.5 °F	0	1	3
		74 °F	1.2	3	1.3
		74.5 °F	3	1.2	0.25

Table 16.14 Revised comfort utility 3 × 3 Strategy—Player 2

met	1.2		40 % RH	50 % RH	60 % RH
clo	0.7		Utility	Utility	Utility
		73.5 °F	0	0.6	1.9
		74 °F	0.75	1.9	1.5
		74.5 °F	2.3	1.4	0.6

Table 16.15 Final comfort utility 3 × 3 Strategy—both players

Player 2 ⟶ / Player 1 ↓	40 % RH	50 % RH	60 % RH
	Utility	Utility	Utility
73.5 °F	0, 0	1, 0.6	**3, 1.9**
74 °F	1.2, 0.75	**3, 1.9**	1.3, 1.5
74.5 °F	**3, 2.3**	1.2, 1.4	0.25, 0.6

Table 16.16   Revised utility 3 $\times$ 3 Strategy—both players

Player 2⟶	40 % RH ($q1$)	50 % RH ($q2$)	60 % RH ($1-q1-q2$)
Player 1↓	Utility	Utility	Utility
73.5 F ($p1$)	0.3, 0.2	0.3, 0.1	0.8, 1
74 F ($p2$)	1, 0	0.8, 0.7	0, 0.5
74.5 F ($1-p1-p2$)	1, 0	0, 0.5	0.5, 1

for Player 2 $= (0.16 \times 0.59 \times 6.2) + (0.16 \times 0.41 \times 7.6) + (0.84 \times 0.59 \times 5.3)$
$+ (0.84 \times 0.41 \times 5) = 5.43$.

Now let us expand our experiment to another 3×3 strategy game. This time assume we have a HVAC control system that is trying to make the best level of comfort for two people (players) working in a lab which due to the type of experiment in the lab the temperature and humidity control strategies are limited to either of 73.5, 74 and 74.5 F temperatures and either of 40, 50, and 60 % relative humidity options. Also assume the Player 1 has a "met" of 1.4 and "clo" of 0.5 and Player 2 has a "met" of 1.2 and "clo" of 0.7. Let us again use PMV-PPD calculator and similar to the process that is defined in the previous example, calculate the PMV, PPD, and the multiplication of the absolute value of the PMV by PPD as the representors of the level of discomfort and gather the information in Tables 16.7, 16.8, 16.9, and 16.10 above.

Similar to the previous example if we reverse the developed discomfort indicators and calculate its natural logarithms, then we can develop the comfort level tables as presented in Tables 16.11 and 16.12 above.

From Tables 16.11 and 16.12 it can be seen that the lowest utility calculated is 4.65. So in order to make the calculations simpler deduct 4.65 from all the utilities and collect the results in Tables 16.13, 16.14, and finally 16.15.

From Table 16.15 we can find out that there are three pure Nash equilibrium at 74.5 °F and 40 % RH, 74 °F and 50 % RH and 73.5 °F and 60 % RH, in which if either player changes his strategy he will be worse off, and the best pure option is the first one (saddle point). Also we can find out that none of the strategies 1, 2, and 3 of player 1 and none of the strategies 1, 2, and 3 for player 2 are dominated by other strategies.

Now assume due to the personal preference of two people working in this place we can revise the assigned utilities as shown in Table 16.16. Here we do not have any pure strategy. In order to calculate the mixed Nash equilibrium in this condition let us assign probabilities of player 1 and player 2 to select their associated strategies to be positive and equal to $p1$, $p2$, $1-p1-p2$, $q1$, $q2$ and $1-q1-q2$ respectively.

From this table we can derive the possible mix payoff for player 1 as follows:

$$\pi 1 = 0.3\,p1q1 + 0.3\,p1q2 + 0.8p1(1 - q1 - q2) + p2q1 + 0.8\,p2\,q2$$
$$+ (1 - p1 - p2)q1 + 0.5\,(1 - p1 - p2)(1 - q1 - q2)$$

And the mixed payoff for player 2 would be as follows:

$$\pi 2 = 0.2\,p1q1 + 0.1\,p1q2 + p1\,(1 - q1 - q2) + 0.7\,p2\,q2$$
$$+ 0.5\,p2\,(1 - q1 - q2) + 0.5\,(1 - p1 - p2)\,q2$$
$$+ (1 - p1 - p2)\,(1 - q1 - q2)$$

Taking partial derivatives of $\pi 1$ with respect to $p1$ and $p2$ and partial derivatives of $\pi 2$ with respect to $q1$ and $q2$ and setting them to zero we would be able to calculate the values for $p1 = 0.55, p2 = 0.42, 1 - p1 - p2 = 0.03, q1 = 0.3, q2 = 0.27$ and $1 - q1 - q2 = 0.43$.

By using these calculated values we can calculate the mixed payoff for player 1 and player 2 as 0.51 and 0.31. Therefore we can set the control of the unit to switch between strategies by the calculated percentage chances for $p1$, $p2$, 1-$p1 - p2$, $q1$, $q2$, and $1 - q1 - q2$.

Another way for calculating the multipliers $p1$, $p2$, $1 - p1 - p2$, $q1$, $q2$, and $1 - q1 - q2$ is to solve the multi-equation set by allowing payoffs of the strategies 1 and 2 and 3 for player 1 to be equal, and similarly the payoffs for strategies 1, 2, and 3 of the player 2 be equal as well. This methods will lead to the same result as the partial derivate method presented above.

References

1. Theodore, L., Turocy T. L., & von Stengel, B. (2001). Game theory. CDAM Research Report LSE-CDAM-2001-09.
2. Saad, W., Han, Z., Poor, V., & Basar, T. (2012). Game theoretic methods for the smart grid. http://arxiv.org/pdf/1202.0452.pdf
3. Schecter, S., Gintis, H. (2012). Game theory in action: An introduction to classical and evolutionary models, Princeton university press, Princeton and Oxford.
4. Forouzandehmehr, N., Perlaza, S. M., Han, Z., & Poor, H. V. (2014). Distributed control of heating, ventilation and air conditioning systems in smart buildings. http://www.Princeton.edu/~Perlaza/Publications/Forouzandehmeher_GLOBECOM_2013.pdf
5. Coogan, S., Ratliff, L. J., Caldrone, D., Tomlin, C., & Sastry, S. (2013). Energy Management via pricing in LQ dynamic games. Department of Electrical Engineering & Computer Sciences at the University of California, Berkeley, 94720, USA.

Part IV
Buildings of the Future

Chapter 17
Buildings of the Future

Abstract The Webster dictionary definition of autonomous is "having the power or right to govern itself." An autonomous building therefore should be a building capable of governing its functions without need for human involvement. Let us dig a little bit deeper and try to explain some of the functions that if a building is capable of governing itself, then it could be eligible to be called autonomous, or in other words smart or intelligent. Searching through different literatures, we can find many diverse definitions for intelligent building all around the world.

Keywords Intelligent building • Buildings of the future • Smart building • Autonomous building • Monitoring • Control • Commissioning • Optimization • Probability distribution • Productivity

The Webster dictionary definition of autonomous is "having the power or right to govern itself". An autonomous building therefore should be a building capable of governing its functions without need for human involvement. Let us dig a little bit deeper and try to explain some of the functions that if a building is capable of governing itself, then it could be eligible to be called autonomous, or in other words smart or intelligent. Searching through different literatures, we can find many diverse definitions for intelligent building all around the world. "In the U.S., an IB (Intelligent Building) is categorized by four basic elements, namely building structure, building systems, building services and building management. In Europe, the emphasis is on information technology and the genuine need of the user. In Singapore and China, it appears that the term "automation" has been dominating with a great emphasis on high technology" [1].

So et al. [1] explain in Japan Intelligent building emphasis is not only on information exchange and supporting management efficiency but also on maintaining an effective working environment, running automatically and comprehensively, and flexibility. Independent from how to look at this topic, it is obvious that the main function of a building from any point of view is to create a safe and comfortable environment in which a specific function or operation has to be performed. Looking back to our main objective in this book which is energy efficiency, it could be derived that from our point of view the building should be capable of performing its functions while conserving the maximum possible energy

© Springer International Publishing Switzerland 2016

J. Khazaii, *Advanced Decision Making for HVAC Engineers*,

DOI 10.1007/978-3-319-33328-1_17

in order to fulfill our expected criteria or its functionality with minimum involvement of humans to be deserved to be named smart or intelligent.

The current advanced buildings typically utilize well-written control programming to manage and monitor the building operation, while relying on commissioning agents and maintenance crew to make proper adjustments to the building systems and building controls during construction and specifically during the building occupancy period based on the collected data from monitoring systems. Generally the control algorithms in response to the relayed signals from the sensors that are located at different locations of the building and its systems, initiate the operation of the actuators that are associated with the HVAC, electrical, and other trade functions to satisfy the space needs, such as temperature, humidity, or lighting level. Examples of such functions are resetting the chilled water temperature set-point in a limited margin, or changing the capacity of outdoor air based on the CO_2 level in each space. In current situation, at best, the information regarding operation of the building systems will be monitored and collected to be reviewed by the commissioning agents or engineering team later on for singling out any deviations from the building and its systems expected behavior. In addition other than parts of the current buildings that their operation is being controlled via utilizing control systems, usually the rest of the current building elements are passive elements that have been designed during the design period, installed during the construction period and will stay unchanged (with the exception of being deteriorated as the time goes by) for the rest of the life of the building. Therefore, the current state of smartness in common buildings can be summarized in rigorous data monitoring and registering in order to give opportunity to the commissioning agents or facility engineers to look into them and detect signs of malfunctioning or irregularities, and fix them manually to keep the desired building functionality while preventing unwanted energy waste, cost, and loss of comfort. Using multiple sensors and real time automatic fault detection are not uncommon where critical applications are required to be managed or very expensive equipment should be protected. But even in such cases it might be too premature to call such functions as the sign of real building intelligence, since for anything to be called smart or intelligent it really should be capable of making smart decisions and behave similar to what is generally associated with being a human with specific skills and capabilities of advanced decision-making.

In the last few chapters in this book what I try to achieve is introducing and briefly describing the most advanced methods of decision-making for designers in order to make the best decisions regarding energy efficiency of the buildings while they are performing different aspects of HVAC design. All the techniques and methods that were represented in the previous chapters are advanced methods that have been proven to break ground in different fronts of scientific fields. For example agent based modeling has made possible an unbelievable advancement in decision-making regarding what can emerge out of simple bottom up actions and reactions of the agents, people or other species. Artificial neural network algorithms on the other hand have made great influence in advancement of pattern recognition and robotics science, while game theory has helped large corporations or even governments to make optimum strategic decisions. My target of writing this book primarily was to draw attention of young HVAC, energy and architectural engineers and students to the possibility of using these advanced methods to improve daily functions of our own field of science parallel to the other scientific fields. That would be great for each of our community of practice members to make best decisions for designing, and other functions that require smart decision-making, but it would be even better if further we could turn our attention from the building designers and what they can do, to the building itself and explore what the building itself could do without intervention of humans. What if the control systems of the building in fact are capable of performing such analyses itself in a routine basis and then improve the system operation accordingly based on the results of its decision-making with some of the presented tools. Wouldn't it be the perfect definition of the buildings of the future, intelligent or smart buildings? Therefore my other and even more important intention has been to point out the vital role that these techniques can play to develop buildings that can think and function autonomously with minimum intervention of human. That I think would be the true definition of buildings of the future, or intelligent and smart buildings.

Accepting this short presented prelude, a real smart building therefore should be capable of performing advanced decision-making algorithms and using the existing data that have been recorded by its monitoring systems to predict what may happen to the building and its systems in the near future, and to operate under the optimal possible conditions to fulfill these building and its systems needs efficiently and cost effectively. Let us look at some of these possible decisions and actions in the following paragraphs.

1. Let's assume buildings of the future are capable of executing programs such as artificial neural network algorithms in their control system and update the building metadata on a continuous basis and therefore make a better and better decision every day and learn how to operate its systems everyday more efficient with higher level of energy saving than past by better recognition of the occupants behavior.
2. Let's assume buildings of the future are capable of using programs such as agent based modeling algorithm and therefore based on the gathered and updated data can predict people behavior who live and work in these buildings and therefore adjust the operation of their systems accordingly. Adjustments such as being able to predict with higher accuracy when the first people gets into the building and when the last one gets out of the building and therefore control system can start the pretreatment of the systems for unoccupied and occupied modes more accurately.
3. Let us assume building control system can perform instant agent based modeling multiple time and during an emergency situation direct the people to the safe by finding the most safe and least crowded path to safety.
4. Assume the buildings of the future can utilize programs such as game theory concepts and improve building control strategies to minimize the cost of keeping maximum zones in the proper temperature range at all time.

5. Assume buildings of the future can use programs such as fuzzy logic to control the comfort level in each space not just based on the room temperature sensor, but also based on a combination of the temperature, relative humidity, metabolisms, level of clouds in the sky and CO_2 level.

6. Assume for each space continuously programs such as agent based modeling can figure out the optimize number of sensors in each instant. The advancement in microelectromechanical sensor technology allows multiple temperature and humidity sensor to control these spaces accordingly without extreme cost.

7. Assume control system can predict the next day electricity consumption based on an updated occupant behavior and therefore plan accordingly to preschedule electricity exchange with a smart grid.

8. Assume microelectromechanical technology (MEMS) allows us to have envelope and glazing systems that are capable of changing their heating characteristics based on the outdoor conditions and Sun position. Also assume the control system can optimize this characteristic so at each instance we will allow only the proper amount of Sun to penetrate to the building and when it is prudent allow internal load to leave the space in order to optimize building energy consumption level.

9. Assume conditions such as where we have refrigerant systems located outside of the building instead of inside of an enclosed space that by code is not required to have refrigerant leak sensors. In such installations for any reason such as small cracks in piping or any kind of sudden impact on refrigerant piping the refrigerant may start to leak out of the unit and into the environment gradually. In this condition and before the maintenance team realizes that the performance of the system has been decreased, a fair quantity of the refrigerant may leak out of the direct expansion units such as air cooled chillers or condensing units associated with the split system air conditioning units. These units are usually dedicated for serving spaces that require all time conditioning (such as communication, audiovisual, and electrical rooms). This is not only undesired condition for the performance of the building but also is extremely undesired for the environment as well. An instant use of algorithms such as artificial neural network in such situations that can develop the normal pattern of energy consumption of the system, and therefore recognizes any major deviation from this normal pattern would help to prevent the energy loss and also the reverse effects on the environment.

10. Let us assume building control system can perform algorithms such as game theory to optimize the control system for the space in order to maintain the best comfort level for different people with different metabolisms and clothing insulations instantly.

11. Let us assume the buildings of the future can utilize decision-making under uncertainty algorithms to provide the probability distribution of its possible energy consumption cost for any time intervals such as 1 or 10 years in advance, have it ready, and therefore give the owners the opportunity of negotiation of this cost with cost guarantors or pre-purchasers, or even change in utilization strategies.

12. Assume in a large campus, each building can utilize game theory algorithm at each time to choose the best overall equilibrium condition for utilizing central plant equipment full or part load operation or type of chiller (absorption or centrifugal) to run at each condition in order to provide the most efficient condition for the whole campus.

It would not be long before we reach the point that we will see autonomous buildings that are capable of evaluating the past data, sensing the existing conditions, and forecasting the future conditions and make the best decisions in order to operate themselves. We as the HVAC and energy engineers should work hard to adopt the best decision-making techniques for the most complicated situations in our design. Whatever we do in our designs now, with no doubt will be autonomously done by building itself in the near future.

Wiggintonet Harris [2] represented many different definitions for intelligent buildings and also intelligent skins that have been used in different literatures since early 1980s—Definitions that emphasize on abilities such as providing responsive, effective, and supportive environment, maximizing efficiency of both occupants and resources, anticipating conditions, taking some initiative in their own operation, flexibility, promoting productivity, predicting the demands, responsivity, and regulating by artificial intelligence.

All of these definitions point to buildings that are capable of collecting and analyzing data and then making decisions autonomously before executing the proper functions.

My hope by writing this book is to point out the available advanced tools and methods that could help the buildings of the future to make proper and intelligent decisions, to the young students and engineers and make them think, be enthusiastic, and try to learn about such advancements in early stages of their careers.

References

1. So, A. T. P., Wong, A. C. W., & Wong, K. C. (2011). A new definition of intelligent buildings for Asia. In *The Intelligent Building Index Manual* (2nd ed.). Hong Kong: Asian Institute of Intelligent Building.
2. Wigginton, M., & Harris, J. (2002). *Intelligent skins*. Oxford: Architectural Press.

Index

Printed in the United States
By Bookmasters